basic electrical
power distribution
VOLUME 1
REVISED SECOND EDITION

basic electrical power distribution
VOLUME 1
REVISED SECOND EDITION

ANTHONY J. PANSINI, E.E., P.E.

Long Island Lighting Company

HAYDEN BOOK COMPANY, INC.
Rochelle Park, New Jersey

To H.M. (Jolly) Jalonack

Friend and Mentor

5   6   7   8   9   PRINTING

75 76 77 78   YEAR

PREFACE

More than a decade has passed since the first edition. It has been a period of progress and change—so substantial, that it is felt a revision of that edition is timely, if indeed not imperative. Althought the basic principles have changed little, if at all, the methods and apparatus used in the delivery of electricity have undergone continuous and impressive improvement—as is to be expected from this progressive industry.

Like the original edition, the text is kept simple and is highly illustrated. Its purpose is to present an overall view of the electric distribution system and its components so that the reader will acquire a firm understanding of how electricity is brought to home and industry.

Grateful acknowledgment is made to the many manufacturers for their contributions of pictures, drawings, and other descriptive material, to Mr. J.M. Barritt for his aid in the preparation of this revision, to Miss June T. Bowe for her stenographic assistance, to the staff of the Hayden Book Company, and finally to my family and friends for their continuing confidence and encouragement.

Anthony J. Pansini

CONTENTS

The Transmission and Distribution System 1

Review 15

Conductor Supports 17

Wood Poles 19

Pole Accessories 30

Review 35

Insulators 37

Conductors 43

Review 53

Line Equipment 55

Distribution Transformers 56

Fuse Cutouts 65

Lightning Arresters 70

Voltage Regulators 75

Capacitors 77

Switches 80

Reclosers 85

Review 87

Obstacles to Overhead Construction 89

Clearances 90

Sag 93

Guying Poles 95

Review 108

Overhead Construction Specifications 110

Index 117

THE TRANSMISSION AND DISTRIBUTION SYSTEM

Electrical Energy—Generator to Home

We are all familiar with pictures of huge dams that store vast quantities of water, and how the water is made to turn the blades of huge turbines, which in turn cause the rotation of electric generators that make electrical energy available for daily use. Whether electrical energy is made available in this manner or simply by turning a generator by means of a gasoline motor and pulley—as is done in an automobile—little purpose is served unless the electrical energy can be delivered to where it can be used.

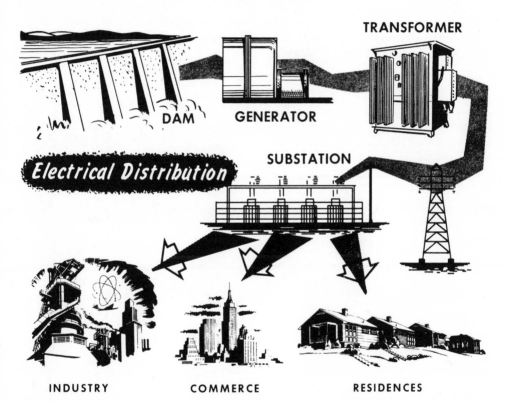

This book deals with the type of equipment, techniques, and problems involved in transmitting electrical energy over long distances from the generator to the device that puts the energy to work, whether it be a large electric motor in a factory or a small electric clock or dishwasher in the home.

As this book develops, the reader will first be made familiar with these pieces of equipment—what they look like and how they are used. Later, he will find a discussion of the electrical principles behind their design and operation.

THE TRANSMISSION AND DISTRIBUTION SYSTEM

The Increasing Use of Electric Service

The increasing widespread use of electric energy may perhaps be attributed to the ease with which it may be converted to other directly usable forms of energy such as light, heat, mechanical energy for operating machinery, X-ray, television and other appliances. Because of its wide application, a utility company must do more than supply electric energy. It must make this energy available at the time and in the manner demanded by the customer. The quality of this energy must be such as to give our lamps proper brilliance, make our motors run at proper speeds and give their full output. What's more, the hazards connected with the distribution of this energy must be eliminated. In short, the utility company must furnish electric *service* as well as electric energy.

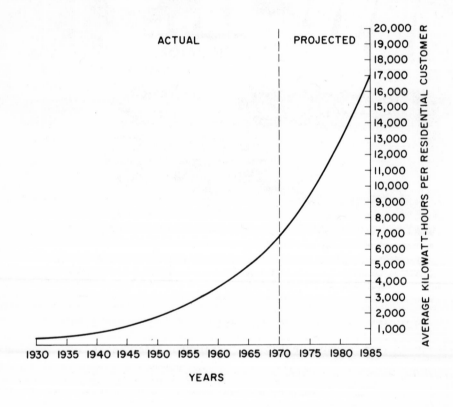

The Average Person Is Using More
Electricity Every Year

Changing Sources of Electrical Energy

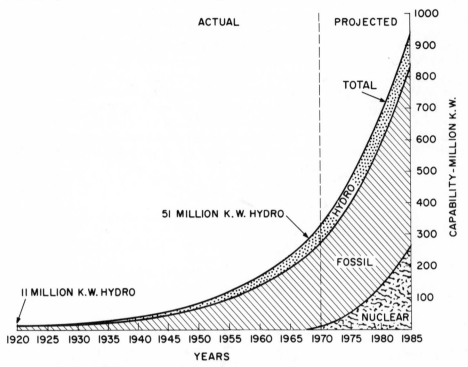

To Meet These
Rising Needs, Electric Capability Must Also Rise

More and More, These Needs Will Be Met by Thermal and Nuclear Energy

Early in this century, as the graph indicates, the chief source of electric energy was water power. Water stored in a dam is potential energy. When scientifically released from its elevated reservoir, water becomes kinetic energy, that is, energy that is capable of doing work.

Hydroelectric power in the United States has grown from a total of almost 11 million installed kilowatts in 1920 to over 50 million kilowatts in 1970. As our graph shows, this tremendous increase in hydroelectric power is really quite small in proportion to the increase in thermal energy. This is partly because of decreased rainfall, but more because of the phenomenal growth of steam power. Today, approximately 13% of the total U.S. power is produced by hydraulic plants; thermal stations employing both fossil and nuclear fuels account for the rest.

Transmission and Distribution

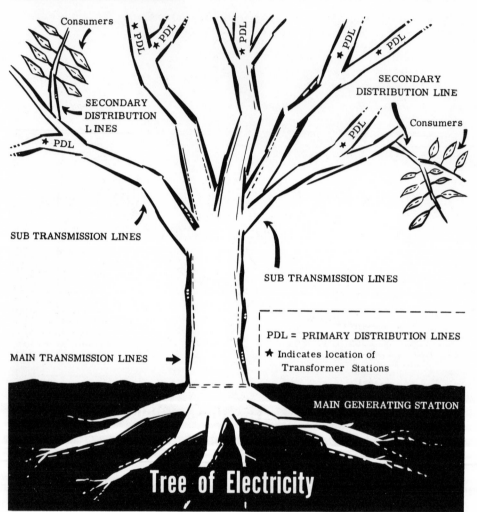

Tree of Electricity

The system of supplying electricity to a community may be compared to a tree, each leaf representing a customer or group of customers. The roots of the tree represent the generating equipment, supplying nourishment to all parts of the tree. The trunk of the tree, which carries all the sap or life of the tree, is similar to the transmission line conducting electricity from the generating station to the various substations in different parts of the system. The branches represent the distributing system conducting electricity to each small branch and leaf.

(1-4)

How the Electric Company Spends its Money

Like any other industry, the electric power system may be thought of as consisting of three main divisions:

1. Manufacture, production or generation.
2. Delivery or distribution.
3. Consumption.

Almost two-thirds of the money invested by the utility company in supplying electric service to its customers is spent in the second of these divisions, distribution. In this book our discussion will be limited to the subject of electric distribution.

Distribution of the Money Invested by the Utility Company In Supplying Electricity

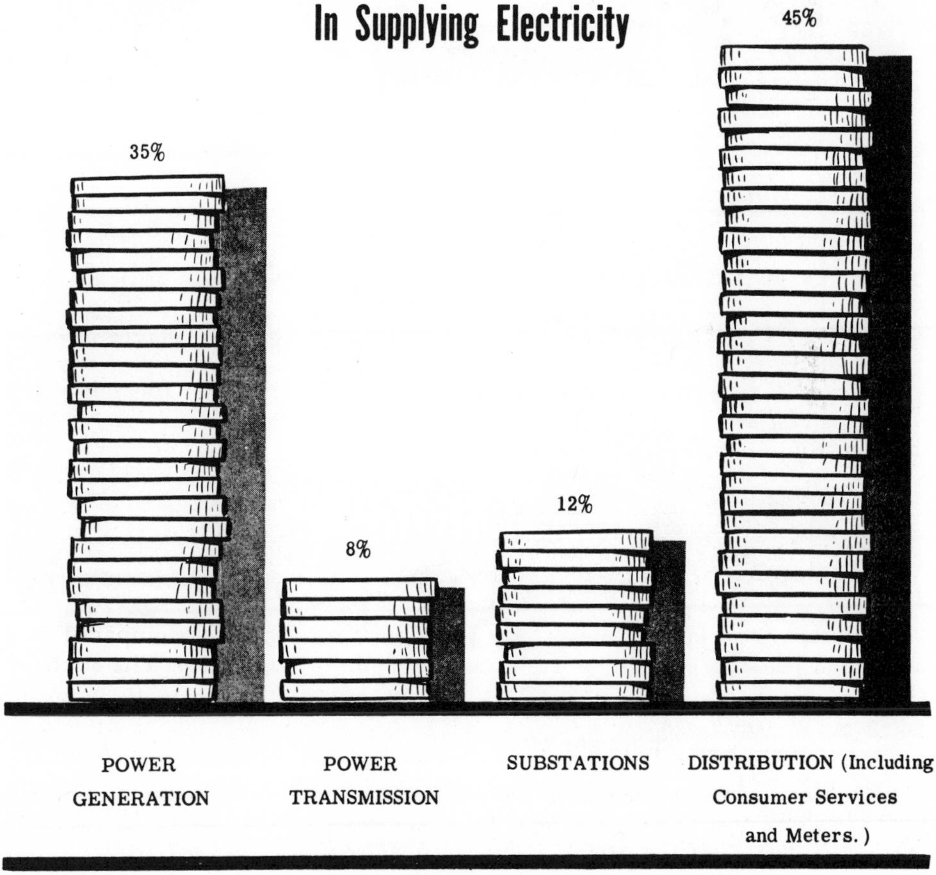

45%

35%

12%

8%

| POWER GENERATION | POWER TRANSMISSION | SUBSTATIONS | DISTRIBUTION (Including Consumer Services and Meters.) |

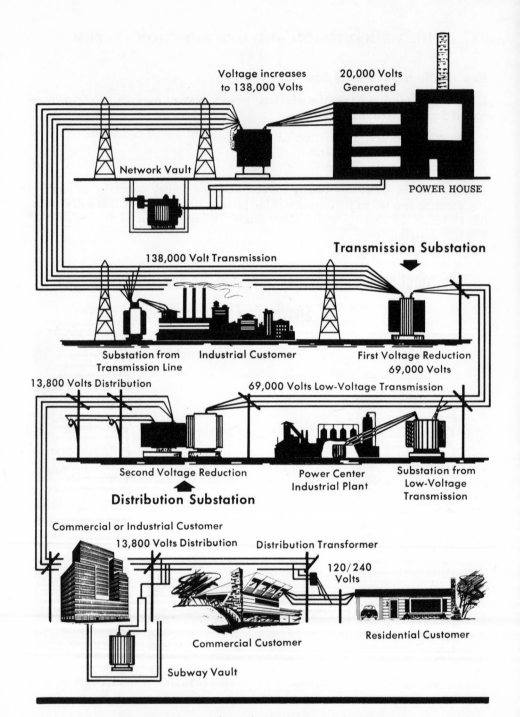

Voltage increases
to 138,000 Volts

20,000 Volts
Generated

Network Vault

POWER HOUSE

Transmission Substation

138,000 Volt Transmission

Substation from
Transmission Line

Industrial Customer

First Voltage Reduction
69,000 Volts

13,800 Volts Distribution

69,000 Volts Low-Voltage Transmission

Second Voltage Reduction

Distribution Substation

Power Center
Industrial Plant

Substation from
Low-Voltage
Transmission

Commercial or Industrial Customer

13,800 Volts Distribution

Distribution Transformer

120/240
Volts

Commercial Customer

Residential Customer

Subway Vault

Electrical Service from the Generator to the Customer

THE TRANSMISSION AND DISTRIBUTION SYSTEM

Power Transmission

On these pages we are showing a typical transmission and distribution system in both pictorial and block diagram forms. Although geographical difficulties, demand variances and other reasons may make for minor differences in some transmission and distribution systems, the voltages chosen here are pretty typical. This is what happens to electricity between the generator and your home, office, store or factory.

There are many definitions of transmission lines, distribution circuits and substations specifying distinctions between them. However, none of these definitions is universally applicable. To give you some idea of where one ends and the other begins, here is a synthesis of definitions accepted by the Federal Power Commission and various state commissions:

> *A transmission system includes all land, conversion structures and equipment at a primary source of supply; lines, switching and conversion stations between a generating or receiving point and the entrance to a distribution center or wholesale point, all lines and equipment whose primary purpose is to augment, integrate or tie together sources of power supply.*

Most generators get their energy from falling water or steam. Waterfalls or water held back by dams are sources of what is known as hydro-electric energy. Sources of steam include coal, oil, natural gas, and, more recently, uranium or so-called nuclear fuels. The use of this last is expected to grow rapidly in future years. The sun, wind, tides and certain exotic chemicals are all possible future sources of electrical energy, but their applications are still in the experimental stages.

In our pictorial rendition, note that the generator produces 20,000 volts. This, however, is raised to 138,000 volts for the long transmission journey. This power is conducted over 138,000-volt (138kv) transmission lines to switching stations located in the important load areas of the territory. These steel tower or wood frame lines, which constitute the backbone of the transmission system, span fields and rivers in direct cross country routes. When the power reaches the switching stations, it is stepped down to 69,000 volts (69kv) for transmission in smaller quantities to the substations in the local load areas. (In some cases it might be stepped down to 13,800 volts [13.8kv] for direct distribution to local areas.) Transmission circuits of such voltages usually consist of open wires on wood or steel poles and structures in outlying zones (along highways, for example) where this type of construction is practicable.

Other transmission-line installations can provide an interchange of power between two or more neighboring Utility Companies to their mutual advantage. Sometimes, in more densely populated areas, portions of these transmission lines may consist of high-voltage underground systems operating at 69,000 volts, 138,000 volts or 220,000 volts.

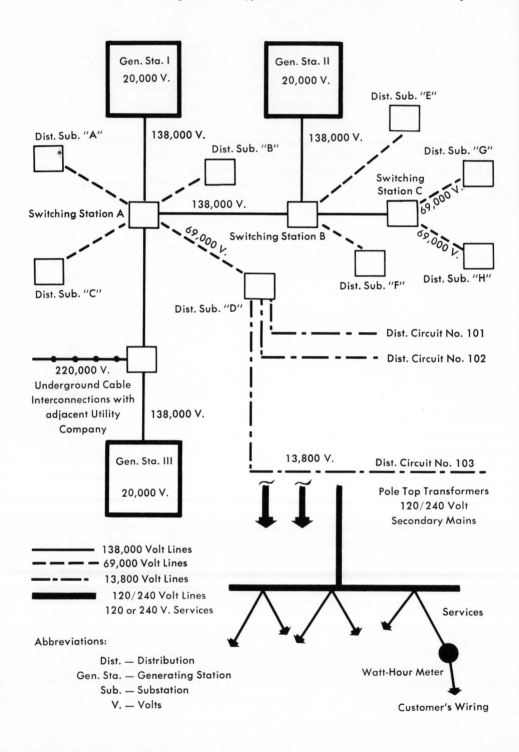

Diagram of a Typical Transmission and Distribution System

THE TRANSMISSION AND DISTRIBUTION SYSTEM

Water-Current Analogy

We have visualized the flow of electric current by comparing it with the flow of water. Where water is conducted in pipes, electric current is made to flow through wires.

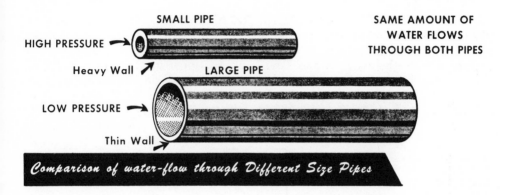

Comparison of water-flow through Different Size Pipes

To move a definite amount of water from one point to another in a given amount of time, we may use either a large-diameter pipe and apply a low pressure on the water to force it through, or we may use a small-diameter pipe and apply a high pressure to the water to force it through. While doing this we must bear in mind that when higher pressures are used the pipes must have thicker walls to withstand that pressure.

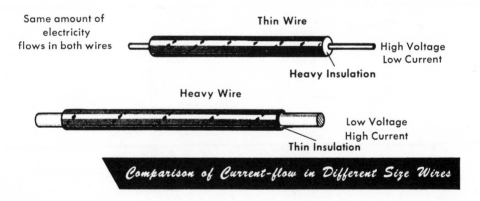

Comparison of Current-flow in Different Size Wires

The same rule applies to the transmission of electric current. In this case, the diameter of the pipe corresponds to the diameter of the wire and the thickness of the pipe walls corresponds to the thickness of the insulation around the wire.

THE TRANSMISSION AND DISTRIBUTION SYSTEM

The Distribution System

At the substations, the incoming power is lowered in voltage for distribution over the local area. Each substation feeds its local load area by means of *primary distribution feeders,* some operating at 2,400 volts and others at 4,160 volts and 13,800 volts.

Ordinarily, primary feeders are one to five miles in length; in rural sections where demands for electricity are relatively light and scattered, they are sometimes as long as 10 or 12 miles. These circuits are usually carried on poles; but in the more densely built-up sections, underground conduits convey the cables, or the cable may be buried directly in the ground.

Diagram of a Typical Distribution System Showing Component Parts

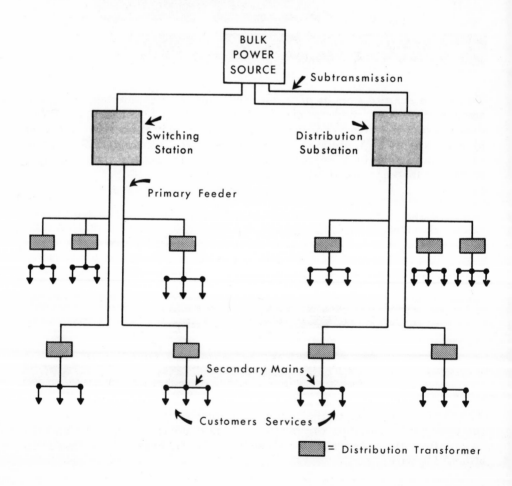

The Distribution System (contd.)

Distribution transformers connect to the primary distribution lines. These transformers step down the primary voltage from 2,400 volts, 4,160 volts, or 13,800 volts, as the case may be, to approximately 120 volts or 240 volts for distribution over secondary mains to the customer's service.

The lines which carry the energy at utilization voltage from the transformer to a customer's services are called secondary distribution mains and may be found overhead or underground. In the case of transformers supplying large amounts of electrical energy to individual customers, no secondary mains are required. Such customers are railroads, large stores and factories. The service wires or cables are connected directly to these transformers. Transformers serve a number of customers and secondary mains; they are located in practically every street in the territory served by utility companies.

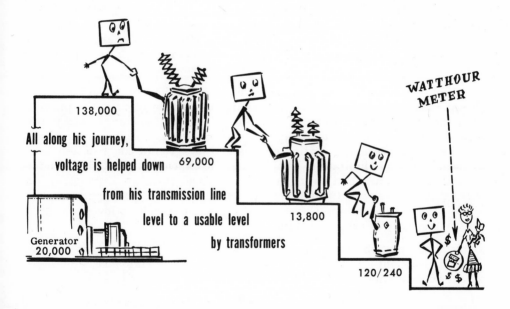

Services and meters link the distribution system and the customer's wiring. Energy is tapped from the secondary mains at the nearest location and carried by the service wires to the customer's building. As it passes on to operate the lights, motors, and various appliances supplied by the house wiring, it is measured by a highly accurate device known as the watt-hour meter. The watt-hour meter represents the cash register of the utility company.

THE TRANSMISSION AND DISTRIBUTION SYSTEM

Determining Distribution Voltages

We pointed out earlier that low voltages require large conductors, high voltages requires smaller conductors. We illustrated this with a water analogy. One can either apply a small amount of pressure and let the water flow through a large pipe or one can apply more pressure and make the water flow through a slimmer pipe. This principle is basic in considering the choice of a voltage (or pressure) for a distribution system.

There are two general ways of transmitting electric current—overhead and underground. In both cases, the conductor may be copper or aluminum, but the insulation in the first instance is usually air, except at the supports (poles or towers) where it may be porcelain or glass. In underground transmission, the conductor is usually insulated with rubber, paper, oil or other material.

In overhead construction, the cost of the copper or aluminum as compared to the insulation is relatively high. Therefore, it is desirable when transmitting large amounts of electric power, to resort to the higher electrical pressures—or voltages, thereby necessitating slimmer, less expensive conductors. Low voltages necessitate heavy conductors which are bulky and expensive to install, as well as intrinsically expensive.

However, there is a limit to how high we can make the voltage and how thin we can make the conductors. In overhead construction there is the problem of supports—poles or towers. If a conductor is made too thin, it will not be able to support itself mechanically. Then the cost of additional supports and pole insulators becomes inordinately high. Underground construction faces the same economic limitation. In this case, the expense is insulation. Underground a cable must be thoroughly insulated and sheathed from corrosion. The higher the voltage, the more insulation is necessary; the bigger the conductor, the more sheathing is necessary. For these reasons, a lower electrical pressure usually is used underground.

Determining distribution voltages is a matter which requires careful studies. Experts work out the system 3 or 4 different ways. For instance, they figure all the expenses involved in a 4000 volt (4 kv) or in a 34.5 kv or a 13 kv system.

The approximate costs of necessary equipment, insulators, switches, etc., and their maintenance and operation must be carefully evaluated. The future with its possibilities of increased demand must also be taken into consideration.

Safety is another important factor. The National Electric Safety Code includes many limitations on a utility company's choice of voltage. Some towns even set up their own standards.

We now see that the utility company must weigh many factors before determining a voltage for distribution.

How Practical Economics Effect the Size of a Transmission Line

Overhead

Heavy conductors -- low voltage -- high current -- long spans --- fewer supports and insulators.

168,000 circular mils

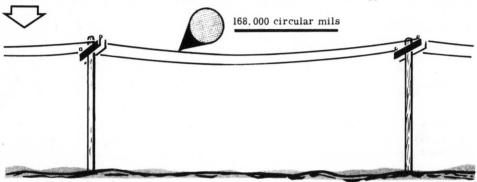

Thin conductors --- high voltage --- shorter spans --- more supports and insulators.

1,620 circular mils

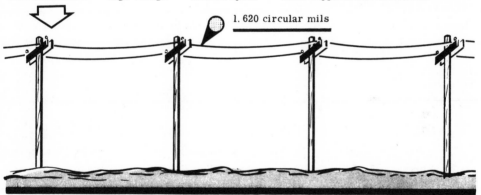

High Voltages require more insulation — more sheathing — lower voltages thus may prove economical

Underground

conductor   insulation   sheath

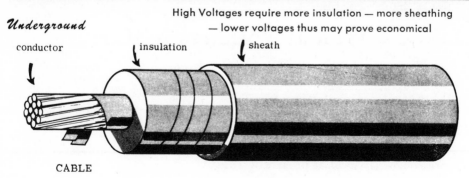

CABLE

Determining Distribution Voltages (contd.)

The Danger of High Voltages

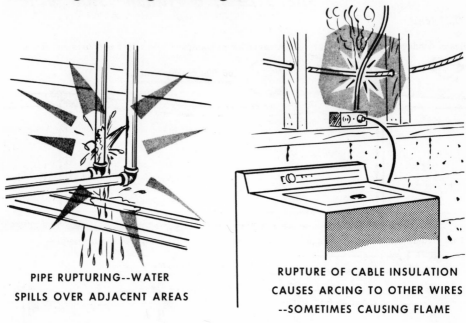

PIPE RUPTURING--WATER
SPILLS OVER ADJACENT AREAS

RUPTURE OF CABLE INSULATION
CAUSES ARCING TO OTHER WIRES
--SOMETIMES CAUSING FLAME

We just mentioned that safety is an important factor in determining voltages for distributing electricity. Here's why! Consider what happens when a water pipe carrying water at high pressure suddenly bursts. The consequences may be fatal and damage considerable. The same is true of electrical conductors. Safeguarding the life and limb of the public as well as workmen is an important responsibility of the utility company.

The table below shows typical transmission and distribution system voltages in use at the present time.

Typical Voltages in Use

Main Transmission	Sub Transmission	Primary Distribution	Distribution Secondary
69,000 V.	13,800 V.	2400 V.	120 V.
*138,000 V.	23,000 V.	4160 V.	*120/240 V.
*220,000 V.	*34,500 V.	*13,800 V.	240 V.
330,000 V.	*69,000 V.		480 V.

*More commonly used at present writing.

The utility company must furnish electric energy in the places, at the times and in the quantities required by consumers. Therefore we say that service is the chief function of the utility company. Besides manufacturing its product, the utility company must distribute energy in a usable form to consumers. Two-thirds of the company's money is spent delivering electricity. A transmission system includes all the equipment necessary to get electric energy from its source of supply to a distribution point. The distribution system includes all the substations, transformers, lines, etc. necssary to get electric energy from the transmission line to the consumer.

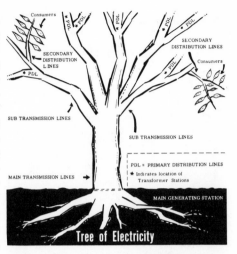

Tree of Electricity

Power is transmitted from the generating station (over transmission lines strung on wood, concrete, or steel poles and structures) to the switching stations. At this point, the voltage is stepped down for transmission to local substations. Each substation feeds its area through primary distribution feeders. The voltage must be stepped down once more before it can be passed through secondary mains to the customer's premises. How much energy the customer taps from the secondary main is measured by the watt-hour-meter.

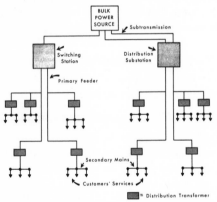

In some ways, transmitting electric energy is like transmitting water. The same amount of water can be moved from one point to another in the same amount of time by using a large-diameter pipe and low pressure—or by using a small-diameter pipe and high pressure. So with electric current. The higher the pressure (voltage) the smaller can be the diameter of the conductor, while low pressure requires a larger conductor.

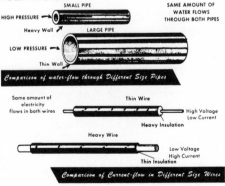

Comparison of water-flow through Different Size Pipes

Comparison of Current-flow in Different Size Wires

REVIEW

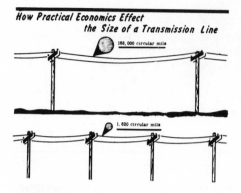

How Practical Economics Effect the Size of a Transmission Line

The pressure of electric energy going through a conductor is called *voltage*. Because conductors (usually copper or aluminum) are expensive, it is theoretically more economical to apply a high voltage across a thin conductor. However, thin conductors need more poles and high voltages can be dangerous. Therefore, practical economics as well as safety sometimes oblige the utility company to use lower voltages than can actually be used.

Throughout the process of transmission and distribution, the voltage (or pressure) of electricity must be raised and lowered (stepped up and stepped down) to meet the particular circumstances. The instrument which does this is called a transformer.

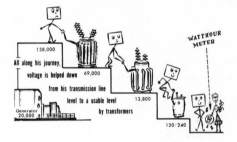

Study Questions

1. What does the electric utility company sell to its customers?
2. What are some sources of electric energy? Which is used more widely today?
3. What are the three main divisions of an electric power system?
4. Distinguish between transmission and distribution.
5. What is the function of a substation?
6. What are the links between the utility company's facilities and the consumer's premises?
7. What are the two ways of distributing electric energy?
8. Compare the flow of water in a pipe with that of electric current in a wire. What is the relationship between water pressure and voltage? Pipe diameter and conductor diameter? Thickness of pipe and insulation?
9. What is electrical pressure called?
10. What are the two most important factors to be considered in determining a distribution voltage?

CONDUCTOR SUPPORTS

Supports for Overhead Construction

Conductors need supports to get from one place to another. Supports may be towers, poles, or other structures. The latter may be made of steel, concrete or wood. The choice of a type of support depends on the terrain to be crossed and the size of conductors and equipment to be carried. Availability and economy, as well as atmospheric elements determine the choice of material.

Usually steel towers are used for transmission lines and wood and concrete poles for distribution circuits. This distinction doesn't always hold true. To meet the needs of a particular circumstance, wood or concrete poles can be used to carry transmission lines; and in some instances a steel tower might be necessary for a distribution circuit.

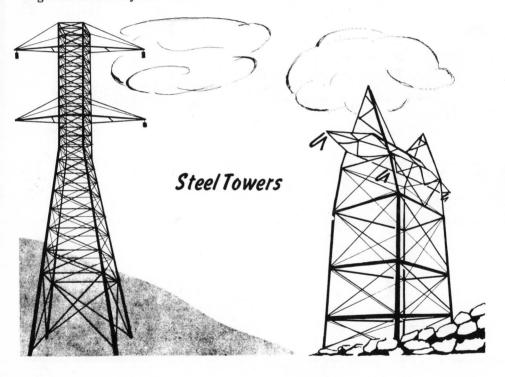

Steel Towers

In general, steel towers are used where exceptional strength and reliability are required. Given proper care, a steel tower is good indefinitely.

Steel can also be used for poles. Although they are comparatively expensive, considerations of strength for large spans, crossing railroads or rivers, for example, make wood undesirable and steel poles, complete with steel crossarms are necessary.

Conductor Supports (contd.)

Reinforced concrete poles are becoming more popular in this country in places where concrete proves more economical. Usually the concrete is reinforced with steel, although iron mesh and aluminum can also be used.

Wood Poles

From Forest

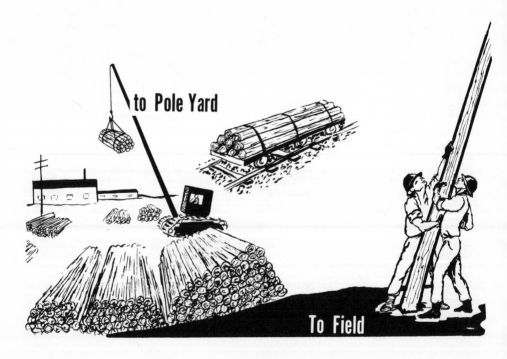

to Pole Yard

To Field

Wood Poles

In the United States, overhead construction more often consists of wood poles with the conductor wires attached to insulators supported on wood crossarms. Although steel and concrete poles are also used, wood has two desirable advantages: initial economy and natural insulating qualities.

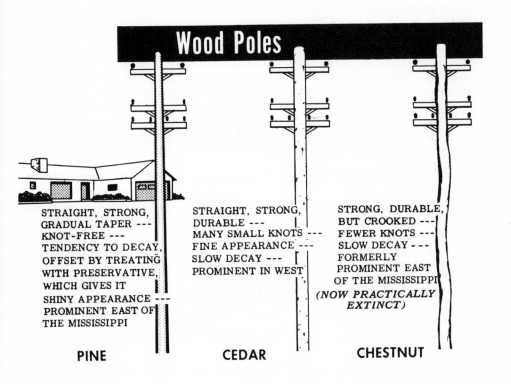

Wood Poles

PINE

STRAIGHT, STRONG,
GRADUAL TAPER ---
KNOT-FREE ---
TENDENCY TO DECAY,
OFFSET BY TREATING
WITH PRESERVATIVE,
WHICH GIVES IT
SHINY APPEARANCE ---
PROMINENT EAST OF
THE MISSISSIPPI

CEDAR

STRAIGHT, STRONG,
DURABLE ---
MANY SMALL KNOTS ---
FINE APPEARANCE ---
SLOW DECAY ---
PROMINENT IN WEST

CHESTNUT

STRONG, DURABLE,
BUT CROOKED ---
FEWER KNOTS ---
SLOW DECAY ---
FORMERLY
PROMINENT EAST
OF THE MISSISSIPPI
(NOW PRACTICALLY
EXTINCT)

The choice of wood for poles depends upon what is available in the particular section of the country. For example, in the central United States, you are most apt to find poles of northern white cedar because it is easily available in Minnesota, Wisconsin and Michigan. Because of the preponderance of western red cedar in Washington, Oregon and Idaho, poles of this wood are found on the Pacific coast. Besides cedar, poles are also made of chestnut or yellow pine. The latter type predominates in the south and east.

WOOD POLES

Kinds of Wood Poles

Cedar is one of the most durable woods in this country. It is light, strong, has a gradual taper and is fairly straight, although full of small knots. Before pine poles came into prominence, cedar poles were used where fine appearance in line construction was required. Chestnut is an extremely strong, durable wood and it is not quite as full of knots as cedar, however, chestnut tends to be crooked. The popularity of both cedar and chestnut in past years was largely attributable to their slow rate of decay, particularly at the ground line. The continuous presence of moisture and air, and the chemicals in the soil tend to encourage mouldy growths which consume the soft inner fibers of wood. This is partially offset by treating the butt (that portion of the pole buried in the ground) with a preservative. Despite the long life of chestnut poles, they are now practically extinct.

Long leaf southern yellow pine is very strong, straight, has a gradual taper and is usually fairly free of knots. Despite its excellent appearance, the use of pine in the past was limited by its lack of durability. However, with improvements in wood-preserving methods, yellow pine has come to be widely used. Laminated poles for greater strength are presently being tried, particularly for transmission lines.

Metal poles, towers and structures are subject to rust and corrosion and, hence, must be maintained (painted, parts renewed) periodically. Wood poles and structures decay and are affected by birds and insects, linemen's climbers, weather, etc., all of which tend to affect their strength and appearance. To combat decay, poles are inspected frequently and treated with preservatives.

The question of appearance of such poles and structures is receiving more and more attention. In addition to "streamlining" such installations (as will be discussed later), color is being introduced. For example, wood poles which were formerly brown or black (the "black beauties", oozing creosote), may now be found in green, light blue, tan, or gray colors.

Pole Length

Two factors must be considered in choosing poles: length and strength required. The length of poles depends on the required clearance above the surface of the ground, the number of crossarms to be attached, and other equipment which may be installed. Provision should also be made for future additions of crossarms, transformers or other devices. Poles come in standard lengths ranging from 25 to 90 ft. in 5-foot differences; that is, 25 ft., 30 ft., 35 ft., etc. Special poles above 90 ft. and below 25 ft. are also available.

How Long a Pole Must Be Depends on...

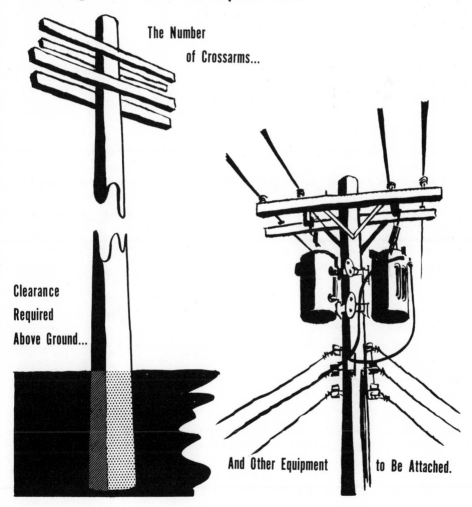

The Number

of Crossarms...

Clearance
Required
Above Ground...

And Other Equipment   to Be Attached.

Pole Strength

Required pole strength is determined by the weight of crossarms, insulators, wires, transformers and other equipment it must carry, as well as by ice and wind loadings. All these forces tend to break a pole at the ground line.

Ice is formed about the conductors and other equipment during a snow or sleet storm. The weight of ice is 57.5 lbs per cubic ft; thus, if ice about 1 inch thick forms about a conductor 100 ft long, more than 100 lbs will be added to the weight carried by the poles. While these direct weights may be appreciable, normal wood poles are more than capable of meeting the ordinary load challenge.

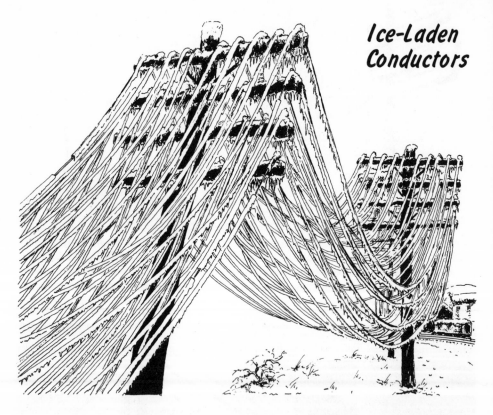

Ice-Laden Conductors

However, the ice formation about the conductors presents quite a surface to the wind. For example, a 60-mile-per-hour wind, blowing against the ice-coated wire mentioned above, will result in a force of more than 135 lb per conductor being applied to the top of the pole. If this pole suspended three conductors the total force would be nearly 400 lb.

Pole Strength (contd.)

It also makes a big difference where the conductor is attached on the pole. A simple illustration of this principle of physics can be seen in your own backyard. If you hang wash on a clothesline attached to the top of a pliable pole, you are not surprised to see the pole bend. To keep the pole from bending, you would attach the line farther down.

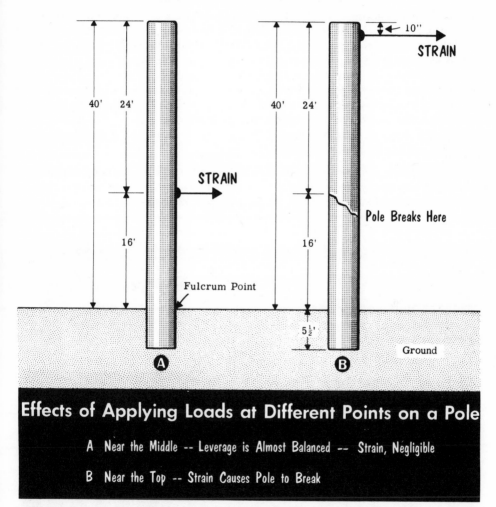

Effects of Applying Loads at Different Points on a Pole

A Near the Middle -- Leverage is Almost Balanced -- Strain, Negligible

B Near the Top -- Strain Causes Pole to Break

The same principle applies to poles which must withstand the strain of wind and ice-laden conductors. The higher above the ground the load is applied, the greater will be the tendency for the pole to break at the ground line.

Pole Strength (contd.)

The forces exerted on a line because of ice and wind will depend on climatic conditions, which vary in different parts of the country. In order to safeguard the public welfare, the U. S. Dept. of Commerce publishes construction standards called the National Electric Safety Code, which divides the United States into three loading districts: heavy, medium and light.

In the heavy loading district, designs of pole lines are based on conductors having a layer of ice .5 inch thick, that is, presenting a surface to the wind of the thickness of the conductor plus 1 inch of ice. Wind pressure is calculated at four lb per sq ft (that of a 60-mile-per-hour wind) and tension on the conductors is calculated at a temperature of 0° (F). In the medium loading district, these values are reduced to .25 inch of ice and a temperature of 15° (F). Wind pressure is calculated at the same four lb per sq ft. In the light loading district, no ice is considered, but a wind pressure of 9 lb per sq ft (that of a wind approximately 67½ miles per hour)—and a temperature of 30° (F) are used for design purposes.

Note that these standards are *minimum*. As an extra precaution, some companies in the heavy and medium districts calculate with a wind pressure of 8 lb per sq ft and some in the light district use 12 lb per sq ft.

Wind and Ice Load Specifications

	Heavy	Medium	Light
Radial thickness of ice (inch)	0.5	0.25	0
Horizontal wind pressure in lb/sq ft	4.	4.	9
Temperature (°F.)	0.	+15	+30

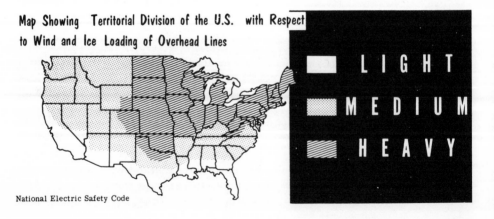

Map Showing Territorial Division of the U.S. with Respect to Wind and Ice Loading of Overhead Lines

LIGHT
MEDIUM
HEAVY

National Electric Safety Code

Pole Strength (contd.)

Another factor which contributes to this bending tendency is the force applied by the wires at the poles. Normally, equal spans of wires are suspended from both sides of a pole. However, should the wire span on one side break, or should there be more wire in the span on one side than on the other, then the uneven pulls will tend to pull the pole over, again giving the pole a tendency to break at the ground line. These uneven pulls are counteracted by guys which we will cover later.

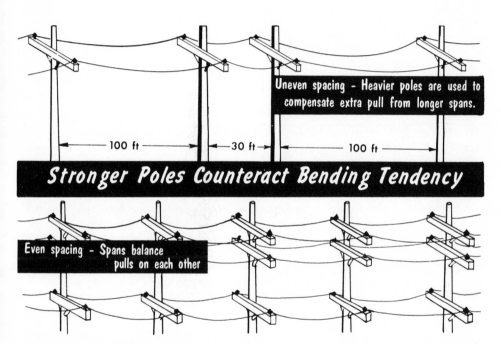

Uneven spacing - Heavier poles are used to compensate extra pull from longer spans.

100 ft — 30 ft — 100 ft

Stronger Poles Counteract Bending Tendency

Even spacing - Spans balance pulls on each other

So we see that although poles may be the same length, they may have different thicknesses at the ground line to give them varying strengths. But it is not enough for a pole to be thick enough at the ground line. If it tapers too rapidly, becoming too thin at the top, then the pole may break at some other point. Therefore, in rating poles for strength, a minimum thickness or circumference is specified not only at the ground line, but also at the top. The strength of a pole is expressed as its *class*. These classes are usually numbered from 1 to 10 inclusive, class 1 being the strongest. Some extra heavy poles may be of class 0, or 00, or even 000. Dimensions are given in the table on the next page.

To describe a pole completely, it is necessary to tell the kind of wood it is made of, its length and its class; for example, 35 ft, class 3, pine pole.

WOOD POLES

STANDARD POLE DIMENSIONS

CLASS			1	2	3	4	5	6	7
Min. top circumference			27"	25"	23"	21"	19"	17"	15"
Min. top diameter			8.6"	8.0"	7.3"	6.7"	6.1"	5.4"	4.8"
Length of Pole	Distance Ground Line to Butt	Kind of Wood	Minimum Circumference at Ground Line (Approximate)						
25	5	P	34.5	32.5	30.0	28.0	26.0	24.0	22.0
		Ch	37.0	34.5	32.5	30.0	28.0	25.5	24.0
		Wc	38.0	35.5	33.0	30.5	28.5	26.0	24.5
30	5½	P	37.5	35.0	32.5	30.0	28.0	26.0	24.0
		Ch	40.0	37.5	35.0	32.5	30.0	28.0	26.0
		Wc	41.0	38.5	35.5	33.0	30.5	28.5	26.5
35	6	P	40.0	37.5	35.0	32.0	30.0	27.5	25.5
		Ch	42.5	40.0	37.5	34.5	32.0	30.0	27.5
		Wc	43.5	41.0	38.0	35.5	32.5	30.5	28.0
40	6	P	42.0	39.5	37.0	34.0	31.5	29.0	27.0
		Ch	45.0	42.5	39.5	36.5	34.0	31.5	29.5
		Wc	46.0	43.5	40.5	37.5	34.5	32.0	30.0
45	6½	P	44.0	41.5	38.5	36.0	33.0	30.5	28.5
		Ch	47.5	44.5	41.5	38.5	36.0	33.0	31.0
		Wc	48.5	45.5	42.5	39.5	36.5	33.5	31.5
50	7	P	46.0	43.0	40.0	37.5	34.5	32.0	29.5
		Ch	49.5	46.5	43.5	40.0	37.5	34.5	32.0
		Wc	50.5	47.5	44.5	41.0	38.0	35.0	32.5
55	7½	P	47.5	44.5	41.5	39.0	36.0	33.5	
		Ch	51.5	48.5	45.0	42.0	39.0	36.0	
		Wc	52.5	49.5	46.0	42.5	39.5	36.5	
60	8	P	49.5	46.0	43.0	40.0	37.0	34.5	
		Ch	53.5	50.0	46.5	43.0	40.0	37.5	
		Wc	54.5	51.0	47.5	44.0	41.0	38.5	
65	8½	P	51.0	47.5	44.5	41.5	38.5		
		Ch	55.0	51.5	48.0	45.0	42.0		
		Wc	56.0	52.5	49.0	45.5	42.5		
70	9	P	52.5	49.0	46.0	42.5	39.5		
		Ch	56.5	53.0	48.5	45.5	43.5		
		Wc	57.5	54.0	50.5	47.0	45.0		
75	9½	P	54.0	50.5	47.0	44.0			
		Ch	59.0	54.0	50.0	47.0			
		Wc	59.5	55.5	52.0	48.5			

P—Long Leaf Yellow Pine
Ch—Chestnut   Wc—Western Cedar

Pole Depth

Soil conditions, the height of the pole, weight and pull factors must be considered in deciding how deep a pole must be planted in the ground. The table below gives approximate setting depths for poles in particular given conditions.

For example, suppose a pole 60 ft long is necessary to clear structures or traffic in the path of the conductors. If there are no extra-strain conditions—for example, the ground is solid, the terrain is flat and the spans are equal—this pole need only be planted 8 ft in the ground. However, if there is an unequal span of wire on one side creating a strain or if the soil conditions are poor, the pole must be set 8½ ft deep.

Typical Pole Setting Depths (approximate)

Length of Pole Overall	Setting Depths on Straight Lines, Curbs and Corners	Setting Depths at Points of Extra Strain or with Poor Soil Conditions
30 ft. & under	5 ft. 0 in.	5 ft. 6 in.
35 ft.	5 ft. 6 in.	6 ft. 0 in.
40 ft.	6 ft. 0 in.	6 ft. 6 in.
45 ft.	6 ft. 6 in.	7 ft. 0 in.
50 ft.	7 ft. 0 in.	7 ft. 6 in.
55 ft.	7 ft. 6 in.	8 ft. 0 in.
60 ft.	8 ft. 0 in.	8 ft. 6 in.
65 ft.	8 ft. 0 in.	8 ft. 6 in.
70 ft. & over	Special Settings Specified	

POLE DEPTH
is determined by

Pole Length   Soil Conditions   Weight and Pull Factors

Pole Gains

Gaining is what we call the process of shaving or cutting a pole to receive the crossarms. In some cases this consists of cutting a slightly concaved recess ½ inch in depth so that the arm cannot rock.

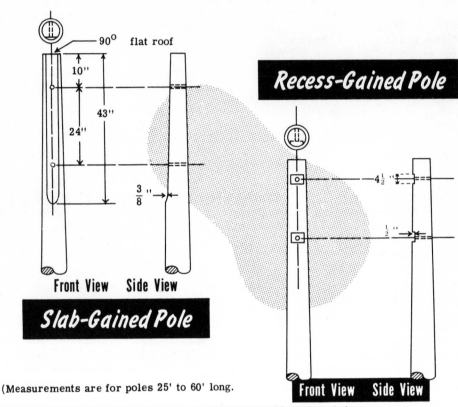

Front View   Side View

Slab-Gained Pole

Recess-Gained Pole

Front View   Side View

(Measurements are for poles 25' to 60' long.

For poles 65' long and over, slab gain should total 73" in length.)

At the present time, the recess in the pole is considered unnecessary. The surface of the pole where the crossarm will be attached is merely flattened by shaving off some of the wood to present a flat smooth area; this is called "slab gaining." The crossarm is then fastened to the pole with a through bolt. Two flat braces are attached to secure the arm. Some poles have two crossarms mounted one on either side of the pole. These are known as double arms. When double arms are installed, the pole is gained on one side only. A double gain would tend to weaken the pole and is unnecessary since the tightening of the through-bolt causes the arm on the back of the pole to bite into its surface.

Pole Roofing

Poles were formerly "roofed" before installation. This roofing at one time consisted of bevels cut from each side to a high point in the middle of a pole, forming a roof similar in shape to the gable roof of a house, or cut at an angle to prevent water, snow or ice from standing on top of the pole and causing decay. Because methods of impregnating poles with preservatives have improved, roofing is now often unnecessary.

POLES ARE 'ROOFED'. . .

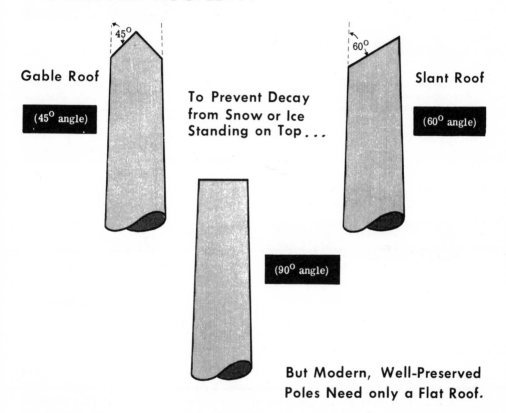

Gable Roof

(45° angle)

To Prevent Decay from Snow or Ice Standing on Top . . .

Slant Roof

(60° angle)

(90° angle)

But Modern, Well-Preserved Poles Need only a Flat Roof.

The gable roof is actually two 45° angles cut on the top of the pole. One common way men have of determining this angle is to measure along the pole a distance equal to ½ of the pole-top diameter. Then they saw from this point to the middle of the pole-top.

Pole steps for the lineman to climb are usually installed at the same time that the pole is roofed and gained, where these are considered to be desirable.

POLE ACCESSORIES

Crossarms

The woods most commonly used for crossarms are Douglas fir or longleaf southern pine because of their straight grain and durability. The top surface of the arm is rounded off so that rain or melting snow and ice will run off easily.

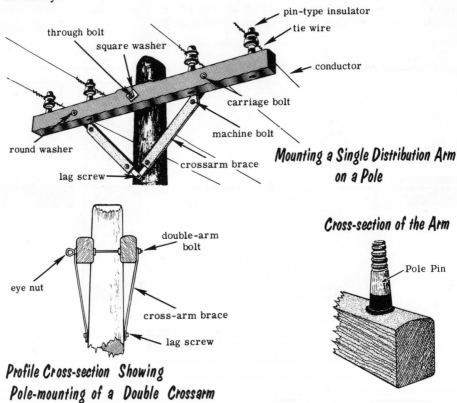

pin-type insulator

tie wire

through bolt

square washer

conductor

carriage bolt

machine bolt

round washer

crossarm brace

lag screw

Mounting a Single Distribution Arm on a Pole

double-arm bolt

eye nut

cross-arm brace

lag screw

Profile Cross-section Showing Pole-mounting of a Double Crossarm

Cross-section of the Arm

Pole Pin

The usual cross-sectional dimensions for distribution crossarms, are $3\frac{1}{2}$ inches by $4\frac{1}{2}$ inches, their length depending on the number and spacing of the pins. Heavier arms of varying lengths are used for special purposes, usually for holding the heavier transmission conductors and insulators. Four-pin, 6-pin and 8-pin arms are standard for distribution crossarms, the 6-pin arm being the most common. Where unusually heavy loading is encountered, as at corner or junction poles, double arms, that is, one on each side of the pole may be required. Again, the emphasis on appearance is causing construction designs to eliminate the installation of crossarms, as will be discussed later.

Pole Pins

Pole pins are attached to the crossarms. They are used to hold pin-type insulators. Note that they are threaded so that the insulator (which we discuss in a few pages) can be securely screwed on.

Pole Pins for Attaching Pin Insulators to Crossarms

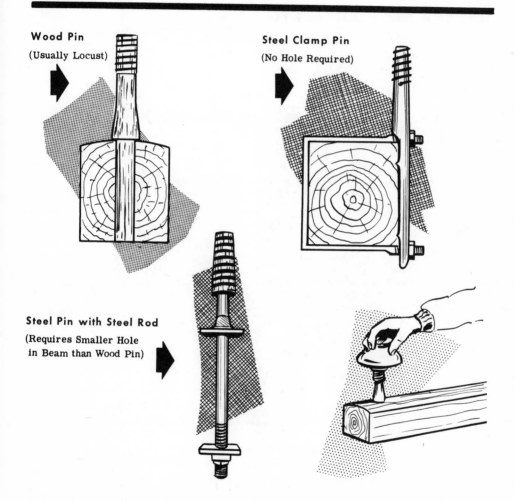

Wood Pin
(Usually Locust)

Steel Clamp Pin
(No Hole Required)

Steel Pin with Steel Rod
(Requires Smaller Hole
in Beam than Wood Pin)

Yellow or black locust wood is most commonly used for attaching the insulators to the crossarms because of its strength and durability, steel pins are used where greater strength is required.

Pin Spacing

The spacing of the pins on the crossarms must be such as to provide enough air space between the conductors to prevent the electric current from jumping or flashing over from one conductor to another. Also, sufficient spacing is necessary to prevent contact between the wires at locations between poles when the wires sway in the wind. In addition, enough space must be pro-

Three 6-Pin Arms Mounted on a Pole

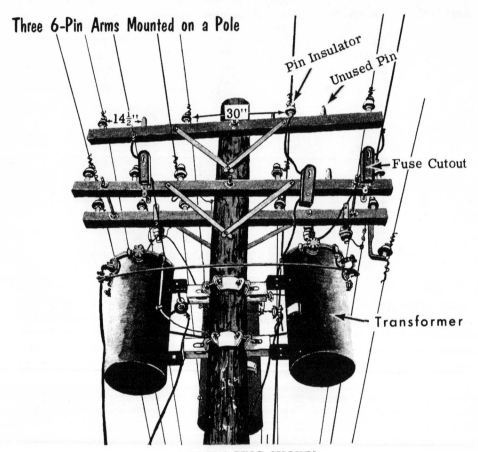

PIN SPACING SHOWN

vided to enable linemen climbing through the wires to work safely. The spacing on a standard six-pin arm is 14½ inches, with 30 inches between the first pins on either side of the pole for climbing space. A special 6-pin arm with spacing wider than 30 inches is frequently used for junction poles to provide greater safety for the lineman.

POLE ACCESSORIES

Secondary Cable

Economy and appearance have dictated the use of wires twisted into a cable for use as secondary mains and for service connections to buildings. The wires carrying current are generally insulated with a plastic material; the neutral conductor is very often left bare and may act as the supporting wire for the cable. The twisted combination, or bundle, is strung from pole to pole as a secondary main, or from pole to building as a service drop or connection. The bundle may consist of 2 wires (duplex), 3 wires (triplex), or 4 wires (quadruplex).

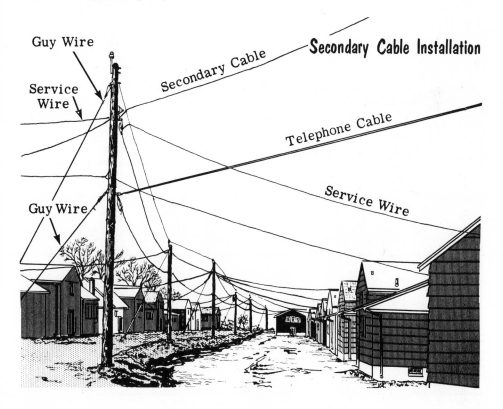

This cable became economical and feasible when manufacturers developed a means of connecting cable to the house service wires without separating the conductors on the pole, as is necessary with secondary racks. These connectors save man-hours and pole space. To install 4 services, the utility company need only install a bracket, a neutral connector and 2 phase connectors. With the secondary rack, *each service* required the installation of a clamp, a hook and 3 connectors.

Secondary Racks

Secondary mains were often supported in a vertical position. When so supported, a so-called secondary rack was used in place of the crossarm. In this type of support, the conductors are spaced closely and are strung on one side of the pole. As the electrical pressure or voltage between these conductors is relatively low, usually 120 or 240 volts, it is not necessary to maintain the same spacing as on crossarms.

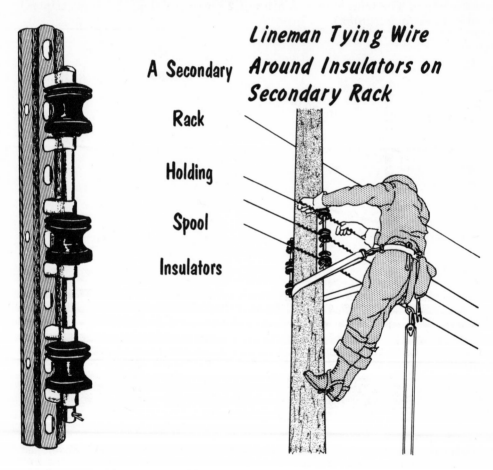

A Secondary Rack Holding Spool Insulators

Lineman Tying Wire Around Insulators on Secondary Rack

The use of this secondary rack simplifies the installation of service wires to the consumer's premises. When a number of services run from each side of the pole a second rack is installed on the opposite side for support of the services on that side.

REVIEW

Conductors carried overhead may be supported on steel towers, steel poles or concrete poles. However, the most commonly used form of overhead construction consists of wood poles with the conductor wires attached to insulators supported on wood crossarms. Cedar and pine poles are most widely used in the U.S. because of their availability and economy. New methods of preservation have made pine most desirable.

POLE DEPTH is determined by

Pole Length Soil Conditions Weight and Pull Factors

The length of a pole is determined by how much equipment must be attached as well as how much clearance above ground is required. Poles come in lengths ranging from 25 ft to 90 ft in 5 ft differences. How deep a pole is set in the ground is determined by its length, soil conditions and weight and pull factors.

A pole must also be strong enough to bear the weight of crossarms, transformers and insulators as well as wires —even when the wires are weighed down by ice and wind. Poles are "classed" according to their strength, a No. 1 pole being the strongest ordinarily used. Therefore to describe a pole completely, specify length, class and wood.

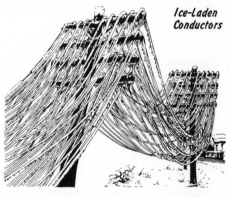

Ice-Laden Conductors

Map Showing Territorial Division of the U.S. with Respect to Wind and Ice Loading of Overhead Lines

Wind and ice loadings depend on varying climactic conditions. To help the utility companies with line design and to safeguard public safety, the Dept. of Commerce has issued construction standards which divide the United States into three wind-and-ice loading areas: heavy, medium and light.

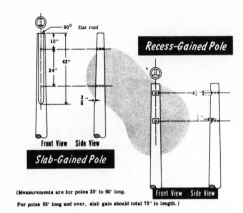

Front View   Side View

Slab-Gained Pole

Recess-Gained Pole

Front View   Side View

(Measurements are for poles 25' to 60' long.

For poles 65' long and over, slab gain should total 73" in length.)

To attach crossarms securely, a pole should be gained. It is roofed to prevent water from staying on top. Attached to the poles may be one or two crossarms, usually having six pins for holding the conductors and insulators. These pins which may be made of steel or wood must be set far enough apart to prevent electrical contact as well as to allow working space.

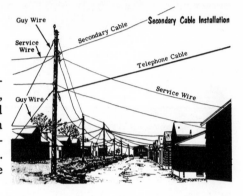

Secondary mains and service connections consist of 2, 3, or 4 wire (duplex, triplex, or quadruplex) cables attached directly to the pole. These have taken the place of racks with spool insulators to which the wires were attached. Economy and better appearance have resulted.

Study Questions

1. What are the major components of overhead construction?

2. Conductor supports are made of three materials. Name them.

3. Of what kinds of wood are poles generally made?

4. How is pole decay reduced to a minimum?

5. What factors should be considered in determining the length of poles to be used in overhead construction?

6. What factors should be considered in determining the strength of the pole required?

7. How is a pole described completely?

8. Define pole *gaining;* pole *roofing.*

9. What are the functions of crossarms and pole pins?

10. What is a secondary rack? What are the advantages of secondary cable?

Insulators

Line conductors are electrically insulated from each other as well as from the pole or tower by nonconductors which we call insulators.

POST TYPE
INSULATOR

PIN TYPE
INSULATOR

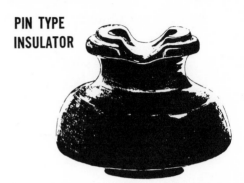

To determine whether or not an insulator can be used, both its mechanical strength and electrical properties must be considered. The most practical insulator materials are porcelain and glass. Both of these leave much to be desired. Porcelain can withstand heavy loading in compression, but tears apart easily under tension—that is when pulled apart. In using a porcelain insulator, therefore, we must take care to make the forces acting on it *compress* and not pull apart. The same is generally true of glass.

Although glass insulators are good for lower-voltage applications, porcelain insulators are much more widely used because they are more practical. Porcelain has two advantages over glass: (1) it can withstand greater differences in temperature—that is, it will not crack when subjected to very high or very low temperatures; (2) porcelain is not as brittle as glass and will not break as easily in handling or during installation.

HOW PORCELAIN REACTS TO COMPRESSION AND TENSION FORCES

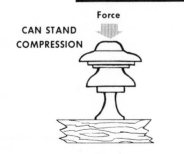

CAN STAND
COMPRESSION

Force

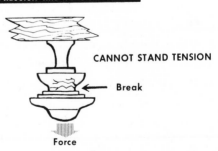

CANNOT STAND TENSION

Break

Force

INSULATORS

Pin-type Insulators

The pin-type insulator is designed to be mounted on a pin which in turn is installed on the crossarm of the pole. The insulator is screwed on the pin and the electrical conductor is mounted on the insulator. Made of porcelain or glass, the pin insulator can weigh anywhere from ½ lb to 90 lb.

TWO STYLES OF LOW-VOLTAGE PORCELAIN INSULATOR

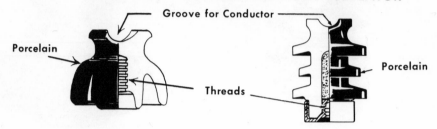

This type of insulator is applicable for rural and urban distribution circuits ... and it is usually constructed as one solid piece of porcelain or glass. In the picture below of a distribution pin insulator, note the grooves for the conductor and for the tie wires.

Larger, stronger pin-type insulators are used for high-voltage transmission lines. These differ in construction in that they consist of two or three pieces of porcelain cemented together. These pieces form what we call petticoats. They are designed to shed rain and sleet easily.

Post type insulators are somewhat similiar to pin type insulators. They are generally used for higher voltage applications, the height and number of petticoats being greater for the higher voltages. They may be mounted horizontally as well as vertically, although their strength is diminished when mounted horizontally. The insulator is made of one piece of porcelain and its mounting bolt or bracket is an integral part of the insulator.

Pin and Post Type Insulators

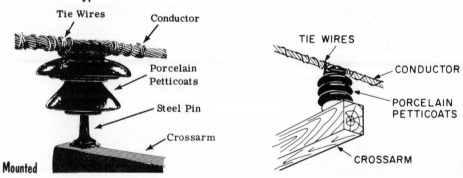

Advantage of Pin or Post Over Suspension Insulators

The most commonly used insulators are the pin or post type and the suspension (or hanging) type. A third type, the strain insulator, is a variation of the suspension insulator, and is designed to sustain extraordinary pulls. Another type, the spool insulator, is used with secondary racks and on service fittings.

The main advantage of the pin or post type insulator is that it is cheaper. Also the pin or post insulator requires a shorter pole to achieve the same conductor clearance above the ground. The pin or post insulator raises the conductor above the crossarm while the suspension insulator suspends it below the crossarm.

Advantage of Pin or Post Insulators over

Suspension Insulators

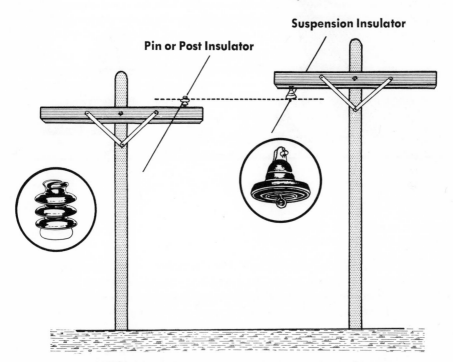

Suspension Insulator

Pin or Post Insulator

Same Conductor Height is Achieved with Shorter Pole.

Suspension Insulators

The higher the voltage, the more insulation is needed. Transmission lines use extremely high voltages, 69,000 to 375,000 volts, for example. At these voltages the pin or post type insulator becomes too bulky and cumbersome to be practical, and the pin which must hold it would have to be inordinately long and large. To meet the problem of insulators for these high voltages, the suspension insulator was developed.

Suspension Insulator - Ball and Socket Type

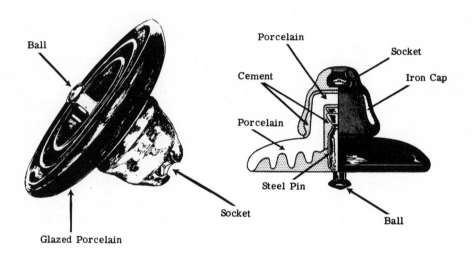

The suspension insulator *hangs* from the crossarm, as opposed to the pin insulator which *sits* on top of it. The line conductor is attached to its lower end. Because there is no pin problem, we can put any distance between the suspension insulator and the conductor just by adding more insulators to the "string."

The entire unit of suspension insulators is called a string. How many insulators this string consists of depends on the voltage, the weather conditions, the type of transmission construction, and the size of insulator used. It is important to note that in a string of suspension insulators one or more insulators can be replaced without replacing the whole string.

INSULATORS

Strain Insulators

Sometimes a line must withstand great strain, for instance at a corner, at a sharp curve, or at a dead-end. In such a circumstance the pull is sustained and insulation is provided by a strain insulator. On a transmission line, this strain insulator often consists of an assembly of suspension insulators. Because of its peculiarly important job, a strain insulator must have considerable strength as well as the necessary electrical properties. Although strain insulators come in many different sizes, they all share the same principle; that is, they are constructed so that the cable will compress (and not pull apart) the porcelain.

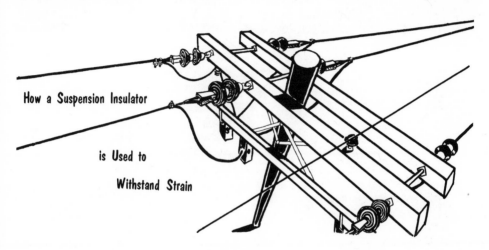

How a Suspension Insulator

is Used to

Withstand Strain

Strain insulators are sometimes used in guy cables, where it is necessary to insulate the lower part of the guy cable from the pole for the safety of people on the ground. This type usually consists of a porcelain piece pierced with two holes at right angles to each other through which the two ends of the guy wires are looped, in such a manner that the porcelain between them is in compression.

STRAIN INSULATOR USED FOR GUY WIRES

Wires of Insulator Pull In Opposite Directions, Resulting in Compression

(1-41)

Spool Insulators

The spool-type insulator, which is easily identified by its shape, is usually used for secondary mains. The spool insulator may be mounted on a secondary rack or in a service clamp. Both the secondary low voltage conductors and the house service wires are attached to the spool insulator. The use of such insulators has decreased greatly since the introduction of cabled secondary and service wires.

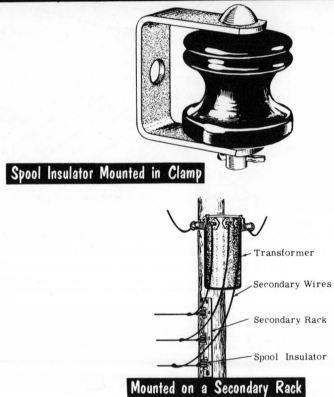

The Spool Insulator May Be Mounted on a Secondary Rack or in a Clamp.

Spool Insulator Mounted in Clamp

Transformer

Secondary Wires

Secondary Rack

Spool Insulator

Mounted on a Secondary Rack

The tapered hole of the spool insulator distributes the load more evenly and minimizes the possibility of breakage when heavily loaded. The "clevis" which is usually inserted in this hole is a piece of steel metal with a pin or bolt passing through the bottom.

CONDUCTORS

Line Conductors

Line conductors may vary in size according to the rated voltage; the number of conductors strung on a pole depends on the type of circuits that are used.

Because they strike a happy combination of conductivity and economy copper, aluminum and steel are the most commonly used conductor materials. Silver is a better conductor than copper; but its mechanical weakness and high cost eliminate it as a possible conductor (during World War II, because of the scarcity of copper, the U.S. Government lent utility companies large amounts of silver to use as substation busbars).

ALL CONDUCTORS ARE COMPARED TO COPPER

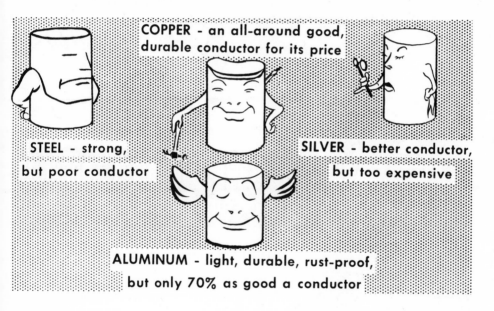

COPPER - an all-around good, durable conductor for its price

STEEL - strong, but poor conductor

SILVER - better conductor, but too expensive

ALUMINUM - light, durable, rust-proof, but only 70% as good a conductor

On the other hand, there are cheaper metals than copper and aluminum; but they would be hopelessly poor conductors. Copper is the touchstone of conductors. Other conducting materials are compared to copper to determine their economic value as electrical conductors.

Aluminum-steel or copper-steel combinations and aluminum have become popular for conductors in particular circumstances. Aluminum alloys are also used as conductors.

CONDUCTORS

Copper Conductors

Copper is used in three forms: hard-drawn, medium-hard-drawn, and soft-drawn (annealed). Hard-drawn copper wire has the greatest strength of the three and is, therefore, mainly used for transmission circuits of long spans (200 ft or more). But its springiness and inflexibility make it hard to work with.

Hard-drawn Copper is Strong but Inflexible and Springy

(used for long spans)

Soft-drawn (annealed) Copper is Weak, but Easy to work With

(used for service and some

distribution circuits)

Soft-drawn wire is the weakest of the three; so its use is limited to short spans and for tying conductors to pin-type insulators. Since it bends easily and is easy to work with, soft-drawn wire is used widely for services to buildings and some distribution circuits. Present practice however, has been toward longer distribution circuit spans and use of medium-hard-drawn copper wire.

Aluminum and ACSR Conductors

The Advantage of Aluminum as a Conductor is its Light Weight - Less than one-third the weight of copper

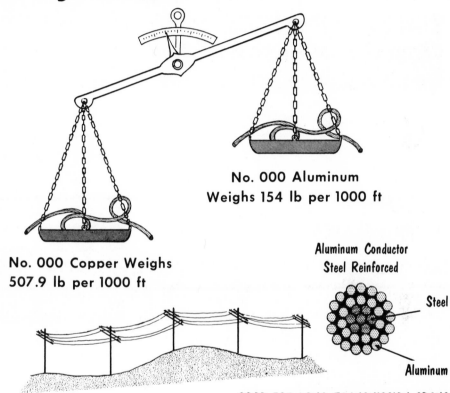

No. 000 Aluminum
Weighs 154 lb per 1000 ft

No. 000 Copper Weighs
507.9 lb per 1000 ft

Aluminum Conductor
Steel Reinforced

Steel

Aluminum

GOOD FOR LONG TRANSMISSION SPANS

Aluminum is also used because of its light weight, which is less than 1/3 that of copper. It is only 60% to 80% as good a conductor as copper and only half as strong as copper. For these reasons it is hardly ever used alone, except for short distribution spans. Usually the aluminum wires are stranded on a core of steel wire. Such steel reinforced aluminum wire has great strength for the weight of the conductor and is especially suitable for long spans. Transmission lines often consist of aluminum conductors steel reinforced (ACSR).

Steel Conductors

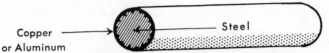

COPPERWELD OR ALUMOWELD CONDUCTOR
Used for Rural Distribution and Guy Wires

Steel by itself is of limited value as a conductor. HOWEVER, it can be successfully combined with copper and aluminum for many conducting uses.

STEEL WIRE
STRONG AND CHEAP,
BUT POOR
CONDUCTOR, RUSTS

ALUMINUM WIRE
VERY LIGHT,
FAIRLY GOOD
CONDUCTOR

COPPER WIRE
EXCELLENT CONDUCTOR,
NOT AS
STRONG AS STEEL

Steel wire is rarely used alone. But where very cheap construction is needed, steel offers an economic advantage. Because steel wire is 3 to 5 times as strong as copper, it permits longer spans and requires fewer supports. However, steel is only about 1/10 as good a conductor as copper and it rusts rapidly. This rusting tendency can be counteracted (so that a steel wire will last longer) by galvanizing, that is, by the application of a coat of zinc to the surface.

Copperweld or Alumoweld Conductors

We have just said that the main disadvantages of steel are a lack of durability and conductivity. On the other hand, steel is cheap, strong and available. These advantages made the development of copper-clad or aluminum-clad steel wire most attractive to the utility companies. To give steel wire the conductivity and durability it needs, a coating of copper is securely applied to its outside. The conductivity of this clad steel wire can be increased by increasing the coating of copper or aluminum. This type of wire, known as Copperweld or Alumoweld is used for guying purposes and as a conductor on rural lines, where lines are long and currents are small.

Conductor Stranding

As conductors become larger, they become too rigid for easy handling. Bending can injure a large solid conductor. For these practical reasons, the stranded conductor was developed. A stranded conductor consists of a group of wires twisted into a single conductor. The more wires in the conductor's cross-section, the greater will be its flexibility. Usually, all the strands are of the same size and same material (copper, aluminum or steel). However, manufacturers do offer stranded conductors combining these metals in different quantities.

TYPES OF CONDUCTORS

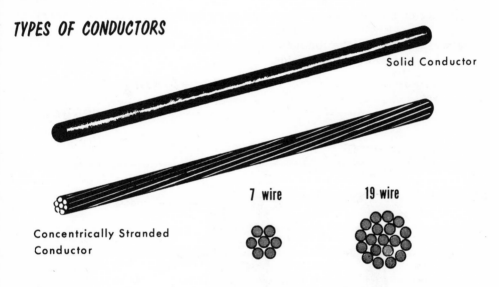

Solid Conductor

Concentrically Stranded
Conductor

7 wire

19 wire

CROSS-SECTION OF STRANDED CONDUCTORS

Sometimes 3 strands of wire are twisted together. But usually, they are grouped concentrically around 1 central strand in groups of 6. For example, a 7-strand conductor consists of 6 strands twisted around 1 central wire. Then 12 strands are laid over those 6 and twisted to make a 19-strand conductor.

To make a 37-strand conductor, 18 more are placed in the gaps between these 18. And so the number of strands increases to 61, 91, 127, etc.

Other combinations are possible. For example, 9 strands can be twisted around a 3-strand twist to make a 12-strand wire. Two large strands can be twisted slightly and then surrounded by 12 twisted strands making a 14-strand conductor.

Conductor Coverings

Conductors on overhead lines may be either bare or covered. Such conductors, located in trees, or adjacent to structures where they may come into occasional contact, may be covered with high density polyethylene or other plastic material resistant to abrasion. This covering is generally not sufficient to withstand the rated voltage at which the line is operating; so these conductors must be mounted on insulators anyway. The purpose of the covering is mainly to protect the wire from mechanical damage. The wires should be treated as though they were bare.

Whether or not the lines are covered, a lineman considers it necessary for safety to work with rubber. Besides wearing rubber gloves and sleeves, he makes sure to cover the conductors, insulators and other apparatus with line hose, hoods, blankets and shields.

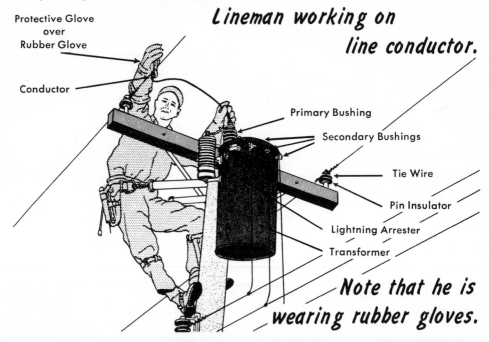

Protective Glove over Rubber Glove

Conductor

Lineman working on line conductor.

Primary Bushing

Secondary Bushings

Tie Wire

Pin Insulator

Lightning Arrester

Transformer

Note that he is wearing rubber gloves.

Recent developments allow the lineman to work on conductors while energized, as long as the platform or bucket in which he is standing is insulated. In this "bare hand" method, he usually wears non-insulating leather work gloves. Extreme caution is necessary in this method, as the insulated platform or bucket insulates the workman from a live conductor and ground, but does not protect him when working on two or more live conductors between which high voltages may exist.

Conductor Coverings (contd.)

Transmission lines, operating at higher voltages, may be worked on when energized or deenergized. They are generally situated in open areas where the danger to the public from fallen wires is negligible. Also, because the amount of insulation required would make the conductor bulky and awkward to install, it is desirable to leave high-voltage transmission line conductors bare.

Wire Sizes

In the United States, it is common practice to indicate wire sizes by gage numbers. The source of these numbers for electrical wire is the American Wire Gage (otherwise known as the Brown & Sharpe Gage). A small wire is

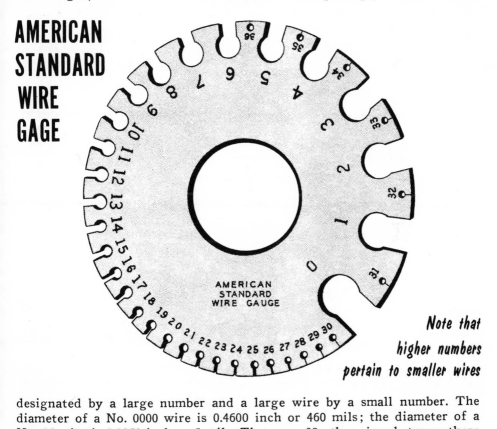

AMERICAN STANDARD WIRE GAGE

AMERICAN STANDARD WIRE GAUGE

Note that higher numbers pertain to smaller wires

designated by a large number and a large wire by a small number. The diameter of a No. 0000 wire is 0.4600 inch or 460 mils; the diameter of a No. 36 wire is 0.0050 inch or 5 mils. There are 38 other sizes between these two extremes. For example, a No. 8 wire is 0.1285 inch (128.5 mils) in diameter and a No. 1 wire is 0.2576 (257.6 mils) in diameter.

CONDUCTORS

Wire Sizes (contd.)

It has proved convenient to discuss the cross-section area of a wire in circular mils. A circular mil (cm) is the area of a circle having a diameter of 0.001 inch or 1 mil. Because it is a *circular area unit* of measure, it is necessary only to square the number of mils given in the diameter of a wire to find the number of circular mils in a circle of that diameter. Thus, a conductor with a 1-mil diameter would have a 1-circular-mil cross-section area; a 3-mil-diameter wire, a 9-cm area; and a 40-mil-diameter wire, a 1600-cm area.

Characteristics of Copper and Aluminum Wire

AWG Size	Diameter (mils)	Cross-section (circular mils)	Resistance (ohms/1000 ft) 20°C Copper	Aluminum	Weight (Lb/1000 ft) Copper	Aluminum
0000	460	211,600	.049	0.080	640.5	195.0
00	365	133,000	.078	0.128	402.8	122.0
0	325	106,000	.098	0.161	319.5	97.0
1	289	83,700	.124	0.203	253.3	76.9
2	258	66,400	.156	0.256	200.9	61.0
3	229	52,600	.197	0.323	159.3	48.4
4	204	41,700	.249	0.408	126.4	38.4
5	182	33,100	.313	0.514	100.2	30.4
6	162	26,300	.395	0.648	79.46	24.1
7	144	20,800	.498	0.817	63.02	19.1
8	128	16,500	.628	1.03	49.98	15.2
9	114	13,100	.792	1.30	39.63	12.0
10	102	10,400	.999	1.64	31.43	9.55
11	91	8,230	1.26	2.07	24.92	7.57
12	81	6,530	1.59	2.61	19.77	6.00
13	72	5,180	2.00	3.59	15.68	4.76
14	64	4,110	2.53	4.14	12.43	3.78
15	57	3,260	3.18	5.22	9.858	2.99
16	51	2,580	4.02	6.59	6.818	2.37
17	45	2,050	5.06	8.31	6.200	1.88
18	40	1,620	6.39	10.5	4.917	1.49
19	36	1,290	8.05	13.2	3.899	1.18
20	32	1,020	10.2	16.7	3.092	0.939
35	5.62	31.5	329.0	540.0	0.0954	0.029
38	4	15.7	660.0	1080.1	0.0476	0.0145

For conductors larger than 0000 (4/0) in size, the wire sizes are expressed in circular mils; for example, 350,000 cm, 500,000 cm, etc. (Sometimes these are expressed as 350 mcm, 500 mcm, etc.)

(1-50)

Tie Wires

Conductors must be held firmly in place on the insulators to keep them from falling or slipping. On pin insulators, they are usually tied to the top or side groove of the insulator by means of a piece of wire, called a tie-wire. On suspension type insulators, conductors are usually held in place by a clamp or "shoe."

Suspension Insulators are
attached to conductor
by a shoe

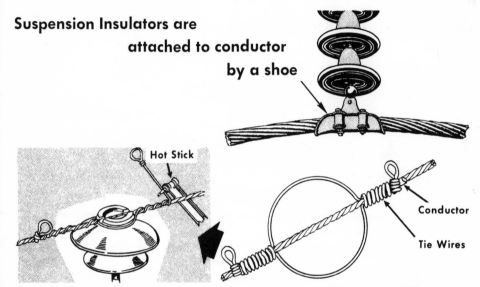

TIE WIRE FOR TOP GROOVE OF PIN INSULATOR (Straightaway)

CLAMP TOP INSULATORS

Where conductors must be maintained while energized, and cannot be touched by hands, they are handled on the ends of sticks called "hot sticks." When it is necessary to use these hot sticks, the ends of the tie wires are looped so that they can be easily wrapped or unwrapped from the insulator.

Connectors

Conductors are sometimes spliced by overlapping the ends and twisting the ends together, taking three or four turns. But to insure a good electrical connection as well as uniformity in workmanship, it is wise to connect conductors with mechanical connectors. They are often substantial money-savers.

Mechanical Connectors

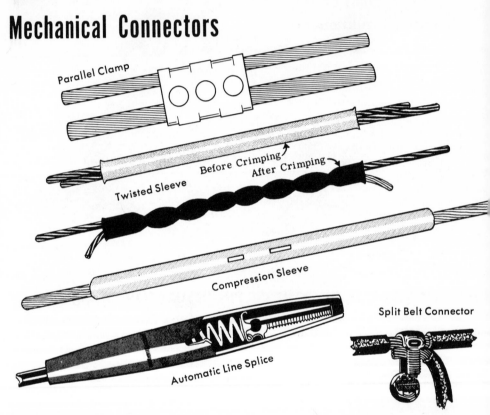

Parallel Clamp

Twisted Sleeve

Before Crimping

After Crimping

Compression Sleeve

Automatic Line Splice

Split Belt Connector

One type inserts the two ends into a double sleeve. When the two conductors are parallel and adjacent to each other, the sleeve is then twisted. With the compression sleeve the conductors are inserted from both ends until they butt and the sleeve is then crimped in several places. The "automatic" splice has the conductor ends inserted in each end where they are gripped by wedges held together by a spring. The split-bolt or "bug nut" connector is a copper or tin-plated-copper bolt with a channel cut into the shank; both conductors fit into the channel and are compressed together by a nut. There are other types of clamps and sleeves; but these represent the most widely used variety.

REVIEW

PIN TYPE INSULATOR

Insulators keep line conductors from making contact with each other or with the pole. Porcelain and glass make good insulators because of their non-conductive natures. Since porcelain is stronger and less fragile, glass is only used for low-voltage installations. Insulators which sit on top of a crossbeam are called pin or post type insulators.

Although the pin insulator is cheaper, there are circumstances which demand the use of the suspension insulator. The latter can be hung in a string. The insulating value of a string can be increased with the addition of more insulators.

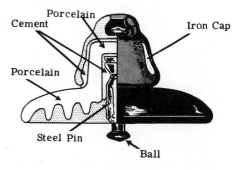

Porcelain

Cement

Iron Cap

Porcelain

Steel Pin

Ball

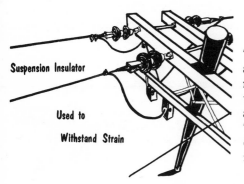

Suspension Insulator

Used to

Withstand Strain

The strain insulator which is used in any situation of exceptional strain may be a small simple affair such as is used on guy wires or it may be a whole assembly of suspension insulators. The spool insulator which has a more distinctive appearance is used primarily on secondary distribution racks.

Copper, aluminum and steel are the most commonly used conductors. Copper is the best conductor. Aluminum is used because of its light weight and steel because of its strength. ACSR (aluminum conductor steel-reinforced) is used for long transmission spans. Copperweld or Alumoweld, a clad-steel combination is used for rural distribution and for guy wires.

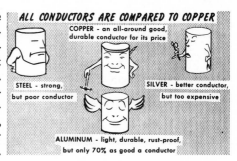

ALL CONDUCTORS ARE COMPARED TO COPPER

COPPER - an all-around good, durable conductor for its price

STEEL - strong, but poor conductor

SILVER - better conductor, but too expensive

ALUMINUM - light, durable, rust-proof, but only 70% as good a conductor

REVIEW

Wire sizes are determined by American Wire Gage numbers. Higher numbers indicate smaller wire sizes and vice-versa. For convenience, the cross-section area of a wire is measured in circular mils (cm).

Note that higher numbers pertain to smaller wires

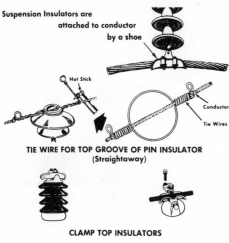

Suspension Insulators are attached to conductor by a shoe

Hot Stick

Conductor

Tie Wires

TIE WIRE FOR TOP GROOVE OF PIN INSULATOR
(Straightaway)

CLAMP TOP INSULATORS

A conductor is firmly attached to a pin or post insulator by a tie wire—to a suspension insulator by a clamp or shoe. Linemen use hot sticks to manipulate tie wires when the lines are energized. Conductors are spliced (joined) by mechanical connectors such as the twisted sleeve, the automatic line splice and the compression sleeve.

Study Questions

1. What is the function of an insulator?

2. Of what materials are insulators for overhead construction usually made? What are their electrical and mechanical properties?

3. What different types of insulators are there?

4. What are the main advantages of pin insulators? Of suspension insulators?

5. Of what materials are overhead line conductors generally made?

6. What characteristics determine a conductor's value?

7. What are the advantages of aluminum as a conductor? Steel?

8. Why are conductors stranded?

9. How are wire sizes expressed?

10. Name several types of mechanical connectors in use.

Line Equipment

Besides conductors and insulators, many other pieces of equipment are necessary to get electric power from the generator to your home. We will want to learn to identify each of these pieces of equipment and know their functions.

1. Distribution Transformers
2. Fuse Cutouts
3. Lightning Arresters
4. Line Voltage Regulators
5. Capacitors
6. Switches (air and oil)
7. Reclosers

LINE EQUIPMENT

Fuse Cutouts

Lightning Arresters

Distribution Transformers

Line Voltage Regulators

Capacitors

Oil Circuit Reclosers

Air-Break Switches

Distribution Transformers

The distribution transformer is certainly the most important of these pieces of equipment. Without the distribution transformer, it would certainly be impossible to distribute power over such long distances. Earlier in this book, we explained that the purpose of a transformer is to step up or step down voltage. In the case of the distribution transformer, the voltage is stepped down from that of the primary mains of a distribution circuit to that of the secondary mains. In most cases, this is from 2400, 4160 or 13,800 volts to 120 or 240 volts.

The Basic Components of the Distribution Transformer

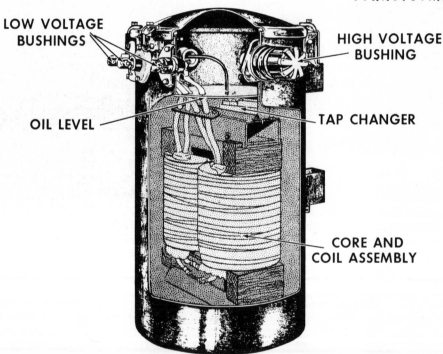

LOW VOLTAGE BUSHINGS

HIGH VOLTAGE BUSHING

OIL LEVEL

TAP CHANGER

CORE AND COIL ASSEMBLY

Most distribution transformers consist of (1) a closed-loop magnetic core upon which are wound two or more separate copper coils, (2) a tank in which the core-coil assembly is immersed in cooling and insulating oil, (3) bushings for bringing the incoming and outgoing leads through the tank or cover.

Bushings

On every distribution transformer, you will notice attachments which we call primary bushings and secondary bushings.

A bushing is an insulating lining for the hole in the transformer tank through which the conductor must pass. Primary bushings are always much larger because the voltage is higher at that point. Sometimes the primary and secondary bushings are called high-voltage and low-voltage bushings.

SIDEWALL MOUNTED BUSHINGS
(Solid Porcelain)

Primary Bushing   Secondary Bushing

Bushings may protrude either from the sidewall of the transformer tank or from its cover. There are three different types of bushings; the solid porcelain bushing, the oil-filled bushing, and the capacitor type bushing.

Solid porcelain bushings are used for voltages up to 15kv. A solid conductor runs through the center of the porcelain form. The conductor is insulated copper cable or solid conductor terminating in a cap. The bushing cap has mountings that permit the line cables to be connected to the transformer winding. The mounting must be so designed that the cable of the transformer lead can be detached to allow the bushing to be removed for replacement.

For higher-voltage transformers (such as we shall see later in our discussion of substations) bushings are oil-filled to improve their insulating characteristics within their specified dimensions. The interior portions of the capacitor bushing are wound with high-grade paper. The oil is replaced with paper and thin layers of metal foil to improve the distribution of stresses because of the high voltage. All three types of bushing have an outer shell of porcelain to contain the insulation inside and to shed rain.

The Tap-Changer

A Tap Changer is Used to Adjust the Turns Ratio of a Transformer

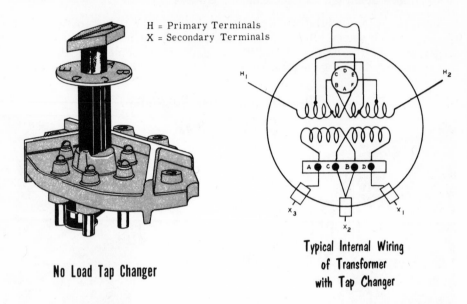

H = Primary Terminals
X = Secondary Terminals

No Load Tap Changer

**Typical Internal Wiring
of Transformer
with Tap Changer**

It is often necessary to vary the voltage in a transformer winding (primary) to allow for a varying voltage drop in the feeder (transmission) lines. In other words, in spite of a varying input, the output must be nearly constant. Several ways can be used to obtain the desired result.

One method used to adjust the winding ratio of the transformer uses the no-load tap changer. A transformer equipped with a no-load tap changer must always be disconnected from the circuit before the ratio adjustment can be made. The selector switch is operated under oil usually placed within the transformer itself; but it is not designed to be used as a circuit breaker. To change taps on small distribution transformers, the cover must be removed and an operating handle is used to make the tap change. For the larger type, one handle may be brought through the cover and the tap may be changed with a wheel or even a motor.

If it is necessary to change the taps when the transformer cannot be disconnected from the circuit, tap changers under-load are used. They involve the use of an autotransformer and an elaborate switching arrangement. The information regarding the switching sequence must be furnished with each transformer. Tap changers can function automatically if designed with additional control circuits: automatic tap changers are used for high-power transformers, and for voltage regulators.

DISTRIBUTION TRANSFORMERS

Mounting Distribution Transformers

Distribution transformers are almost always located outdoors where they are hung from crossarms, mounted on poles directly or placed on platforms. In general, transformers up to 75 kva size are mounted directly to the pole or on a crossarm and larger size transformers (or groups of several transformers) are placed on platforms or mounted on poles in banks or clusters.

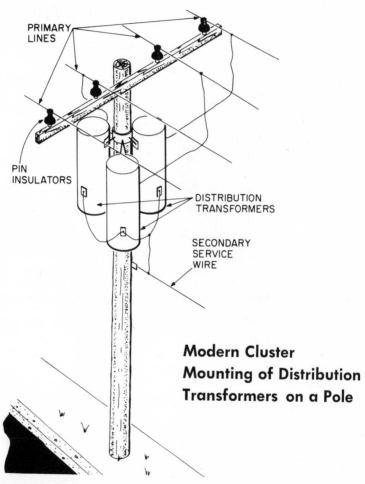

PRIMARY LINES

PIN INSULATORS

DISTRIBUTION TRANSFORMERS

SECONDARY SERVICE WIRE

Modern Cluster Mounting of Distribution Transformers on a Pole

How a transformer is mounted is a matter of considerable importance. Remember that the distribution transformer must stay put and continue functioning even in the midst of violent winds, pouring rain, freezing cold, sleet and snow. Besides weather, there is the danger of the pole itself being hit by a carelessly-driven automobile.

Mounting Distribution Transformers (contd.)

Modern pole-mounted transformers have two lugs welded directly on the case; these lugs engage two bolts on the pole from which the whole apparatus hangs securely. This method, which is known as *direct mounting* eliminates the need for crossarms and hanger irons (as was done in the past), thus saving a considerable amount of material and labor.

Direct Pole-Mounting of Distribution Transformers

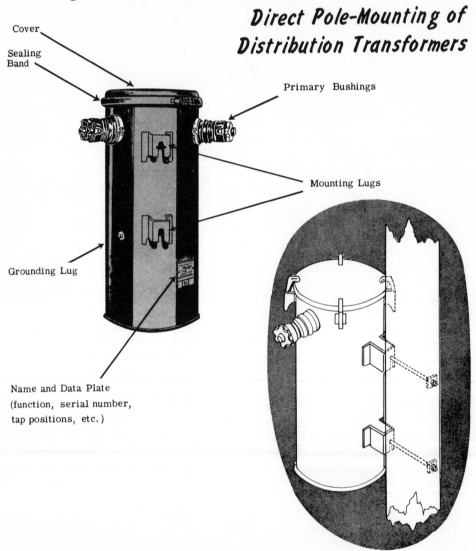

Cover

Sealing Band

Primary Bushings

Mounting Lugs

Grounding Lug

Name and Data Plate
(function, serial number,
tap positions, etc.)

Mounting Distribution Transformers (contd.)

In the past, transformers were hung from crossarms by means of hanger irons, which were two flat pieces of steel with their top ends bent into hooks with squared sides. The transformer was bolted to these pieces of steel, the assembly was raised slightly above the crossarm and then lowered so that the hooks on the hanger irons would engage the crossarm.

Hanger-iron Method of Mounting Distribution Transformers on Poles

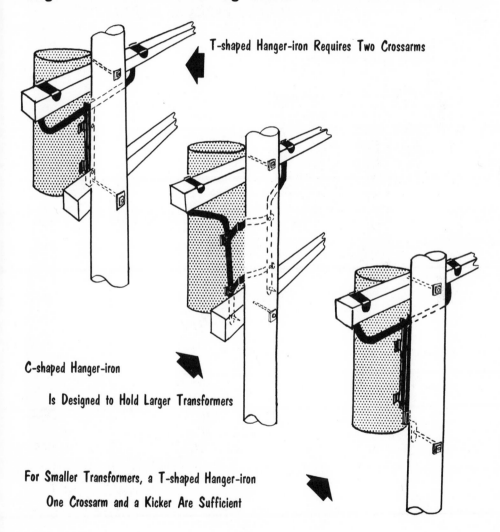

T-shaped Hanger-iron Requires Two Crossarms

C-shaped Hanger-iron

Is Designed to Hold Larger Transformers

For Smaller Transformers, a T-shaped Hanger-iron
One Crossarm and a Kicker Are Sufficient

Mounting Distribution Transformers (contd.)

A transformer should not be mounted on a junction pole (a pole supporting lines from three or more directions) as this makes working on such a
pole more hazardous for the lineman.

Mounting
of Distribution Transformers

TRANSFORMER

PLATFORM

TRANSFORMER
CONTAINED IN
METAL TANK

PAD

PLATFORM MOUNTED   PAD MOUNTED

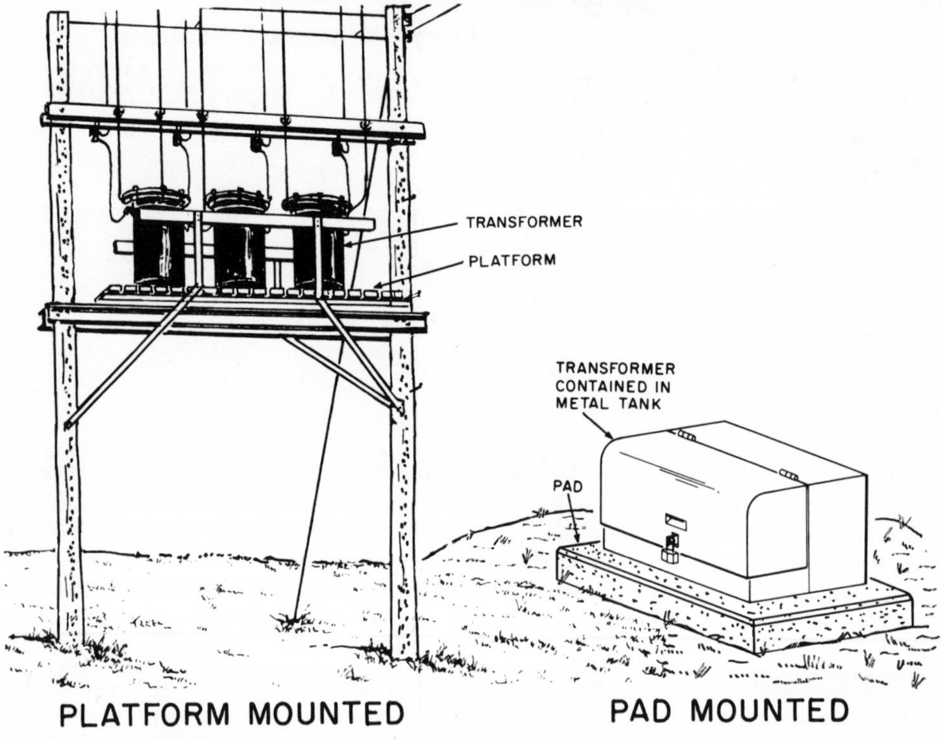

Where transformers cannot be mounted on poles because of size or number,
they may be installed on an elevated platform or a ground-level pad. Platforms are built in any shape or size required to suit the particular need.
They are usually constructed of wood, though steel is often used for some
of the members. Ground-level pads are usually made of concrete or reinforced concrete and have provisions for enclosing the transformers within
a fence or wall or other enclosure for safety reasons. Ground-level pads are
very useful when appearance is a major consideration.

A Conventional Distribution Transformer

A Conventional Transformer Requires Separate Mounting of Lightning Arresters and Fuse Cutouts

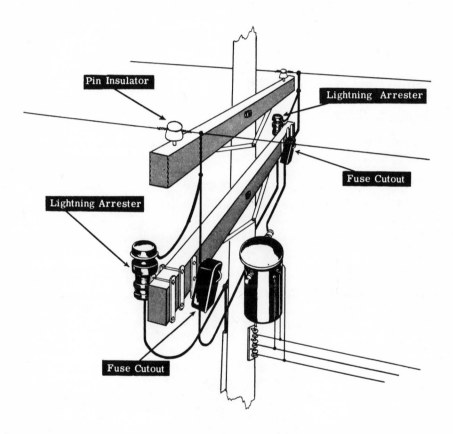

Manufacturers produce two types of distribution transformers: the conventional and the completely-self-protected. (CSP is the Westinghouse trade designation for this latter type. Other manufacturers simply call them self-protected transformers.) A conventional distribution transformer consists only of a case containing the transformer unit. Protective devices, usually a primary fuse cutout and a lightning arrester, are mounted separately on the pole or crossarm.

CSP Distribution Transformers

In the CSP transformer, a *weak link* or primary protective *fuse* link is mounted inside the tank with the transformer unit as also are two circuit breakers for protection on the secondary side of the transformer. A simple thermal device causes the breakers to open when a predetermined safe value of current is exceeded. The lightning arrester is mounted on the outside of the tank. It is apparent that the CSP transformer makes for simpler, more economic mounting and neater appearance. What's more, it is of particular advantage for higher-voltage (13.8 kv) primary distribution systems where connections and disconnections are made with hot sticks.

A 'Self-Protected' Transformer
Lightning Arresters and Circuit Breaker are Integral Parts of its Design

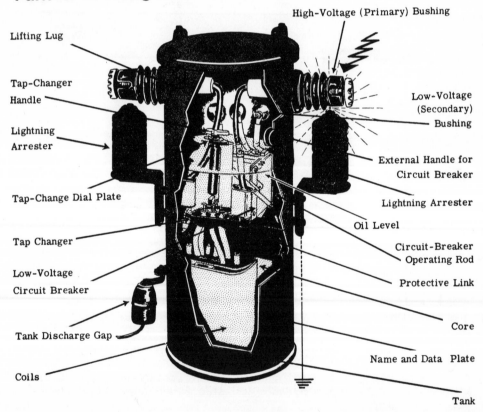

High-Voltage (Primary) Bushing

Lifting Lug

Tap-Changer Handle

Low-Voltage (Secondary) Bushing

Lightning Arrester

External Handle for Circuit Breaker

Tap-Change Dial Plate

Lightning Arrester

Oil Level

Tap Changer

Circuit-Breaker Operating Rod

Low-Voltage Circuit Breaker

Protective Link

Tank Discharge Gap

Core

Coils

Name and Data Plate

Tank

Fuse Cutouts

Suppose a circuit is designed to carry at 100 amps. Should the amperage climb over that limit, it could eventually melt some of the wires and cause widespread interruption of service. To prevent this, we intentionally put a weak spot in a circuit—a place where overload will register and open the circuit almost immediately. We call this spot a fuse. A fuse consists of a short piece of metal having low melting characteristics which will melt at a rated temperature.

It is amperage flowing through a conductor which sometimes makes the conductor hot to touch. This is what the fuse counts on. Should there be an overload, the fuse melts, thus disconnecting the circuit.

Fuse cutouts go one step further. They may be placed so as to cut out the section of the circuit which is endangered, allowing the rest of the circuit to remain energized.

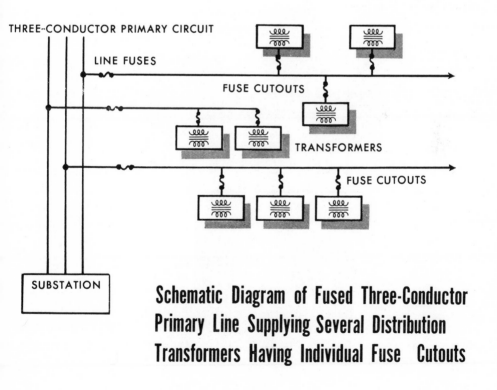

Schematic Diagram of Fused Three-Conductor Primary Line Supplying Several Distribution Transformers Having Individual Fuse Cutouts

(1-65)

Fuse Cutouts (contd.)

A primary fuse cutout is connected between the primary lines and the transformer to protect the transformer from overloads and to disconnect the transformer from the primary lines in case of trouble. What we call a primary line fuse cutout can disconnect any portions of a primary circuit supplying several distribution transformers, in case of overload or fault, leaving the rest of the circuit energized.

Schematic Diagram of a Single-Conductor
Primary Line Employing a Fused Transformer

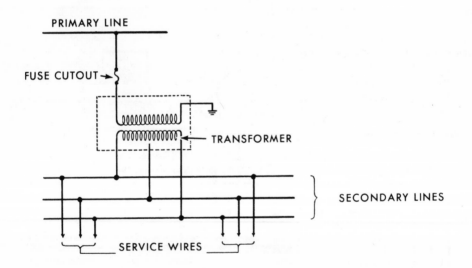

Although there are three main types of fuse cutouts, the principle upon which each is constructed is the same. A fuse ribbon makes a connection between two contacts, either the line and the transformer or the main line and that portion to be protected.

FUSE CUTOUTS

The Door Type Cutout

In the door or box type cutout, the fuse is mounted inside the door in such a manner that when the door closes, the fuse engages two contacts, one on the top of the box and the other on the bottom. To open this cutout, the door is pulled open and allowed to hang downward from the box. The fusible element is enclosed in a fiber tube; when the fuse blows or melts because of

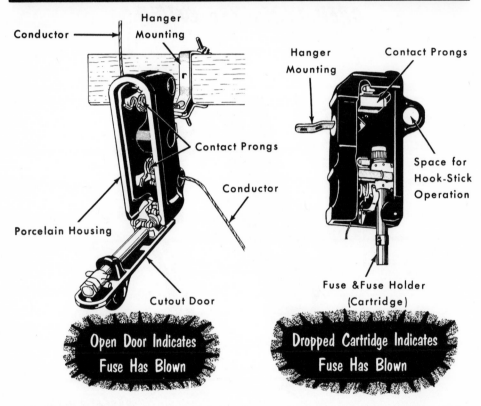

Two Styles of Door Fuse Cutout

Conductor

Hanger Mounting

Contact Prongs

Porcelain Housing

Conductor

Cutout Door

Hanger Mounting

Contact Prongs

Space for Hook-Stick Operation

Fuse & Fuse Holder (Cartridge)

Open Door Indicates Fuse Has Blown

Dropped Cartridge Indicates Fuse Has Blown

excessive current passing through it, the resultant arc attacks the fiber tube, producing a gas which blows out the arc. For this reason this type cutout is sometimes also called the expulsion type. In later models, the fuse mounting is arranged so that the melting of the fusible element causes the door to drop open, signaling to the man on the ground that the fuse has blown. The use of this type cutout is generally limited to circuits operating at voltages under 5000 volts.

FUSE CUTOUTS

The Open-Type Cutout

The open-type cutout is essentially the same as the door type, except that the fiber tube enclosing the fusible element is exposed in the open, rather than enclosed in a porcelain box. This arrangement enables larger currents to be interrupted without confining as much the attendant violent expulsion of gases (which can destroy the cutout). The tube drops when the fuse blows, indicating that the fuse has blown. This type cutout is used on distribution circuits operating at voltages over 5000 volts, though obviously it can be used on lower voltage circuits.

OPEN-TYPE FUSE CUTOUT

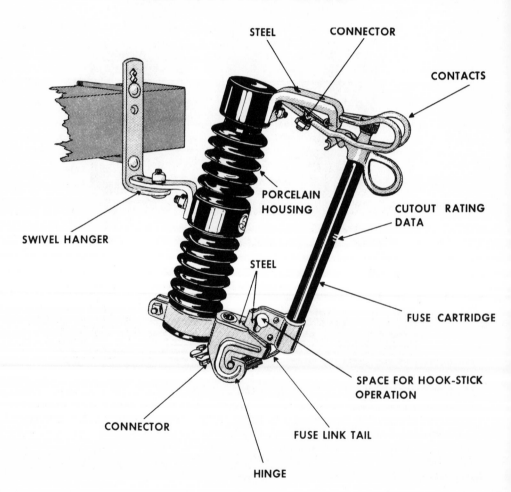

The Repeater Fuse

When line fuses are used to protect a portion of a primary circuit as described above, a repeater fuse may be used. The repeater fuse is usually of the open type and consists of two or three fuses mechanically arranged so that when the first fuse blows and drops, the action places the second fuse automatically in the circuit. If the trouble has been cleared, service will

Repeater-Type Fuse Cutout

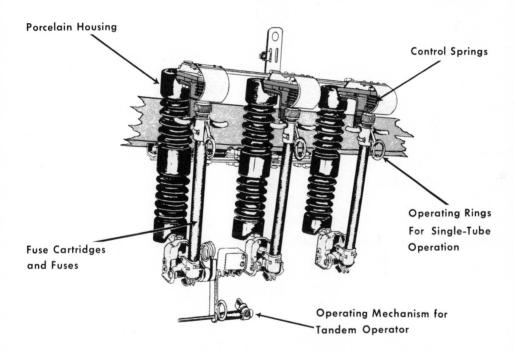

Porcelain Housing

Control Springs

Operating Rings
For Single-Tube
Operation

Fuse Cartridges
and Fuses

Operating Mechanism for
Tandem Operator

be restored. Should the second fuse also blow, a third is also automatically connected in the circuit; when the third fuse blows, the portion of the circuit is finally de-energized. Repeater fuses hold down to a minimum interruptions in service caused by temporary faults. These faults may arise from wires swinging together when improperly sagged, or from tree branches or animals making momentary contact with the line, or from lightning surges causing temporary flashover at an insulator.

All of these fused cutouts are mounted on the crossarm or on the pole by bolts and a steel bracket.

Lightning Arresters

A lightning arrester is a device which protects transformers and other electrical apparatus from voltage surges. These surges can occur either because of lightning or improper switching in the circuit. The lightning arrester provides a path over which the surge can pass to the ground before it has a chance to attack and seriously damage the transformer or other equipment.

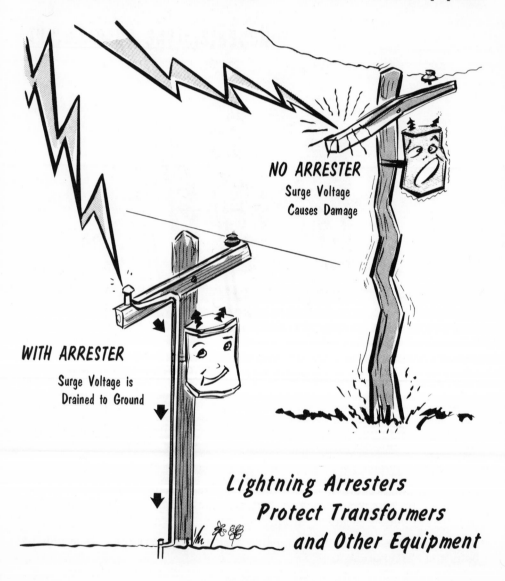

NO ARRESTER
Surge Voltage
Causes Damage

WITH ARRESTER

Surge Voltage is
Drained to Ground

*Lightning Arresters
Protect Transformers
and Other Equipment*

Lightning Arresters (contd.)

The elementary lightning arrester consists of an air gap in series with a resistive element. The voltage surge causes a spark which jumps across the air gap, passes through the resistive element (silicon carbide, for example) which is usually a material which allows a low-resistance path for the high voltage surge, but presents a high resistance to the flow of line energy; this material is usually known as the "valve" element. There are many different

The Elementary Lightning Arrester

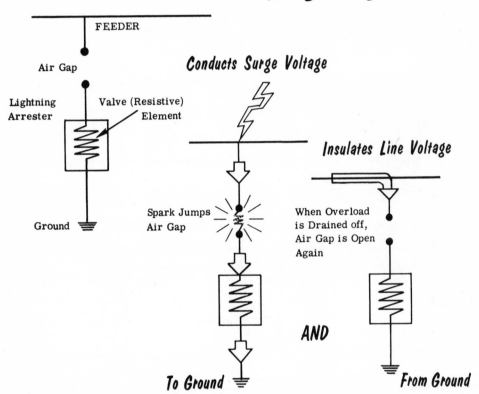

types of lightning arresters, but they generally have this one principle in common. There is usually an air gap in series with a resistive element, and whatever the resistive (or valve) element is made of, it must act as a conductor for high-energy surges and also as an insulator toward the line energy. In other words, the lightning arrester leads off *only* the surge energy. Afterwards, there is no chance of the normal line energy being led into the ground.

LIGHTNING ARRESTERS

Valve Arresters

Since all arresters have a series gap and a resistive element, they differ only in mechanical construction and in the type of resistive element used. One type arrester consists of a porcelain cylinder filled with lead peroxide pellets and has electrodes at either end. The series gap assembly is usually at the top. When there is excessive voltage surge on the line, the spark jumps across the air gap and the surge energy flows through the row of pellets to the ground. As the surge decreases, the resistive power of the pellets increases so that no line energy will flow to the ground. Later type arresters replaced the lead peroxide pellets with other suitable materials.

Valve Lightning Arrester

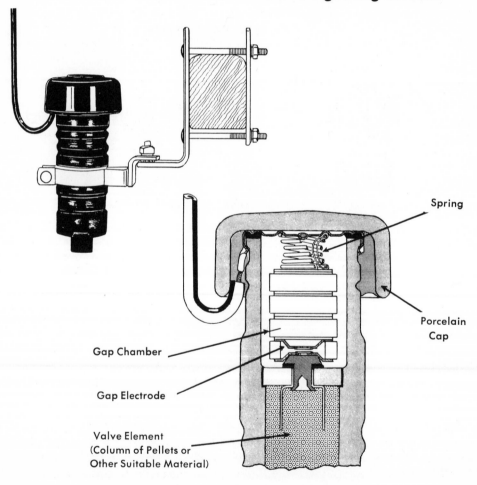

Spring

Porcelain Cap

Gap Chamber

Gap Electrode

Valve Element
(Column of Pellets or
Other Suitable Material)

Arresters with Isolators

Isolator Feature on Lightning Arrester

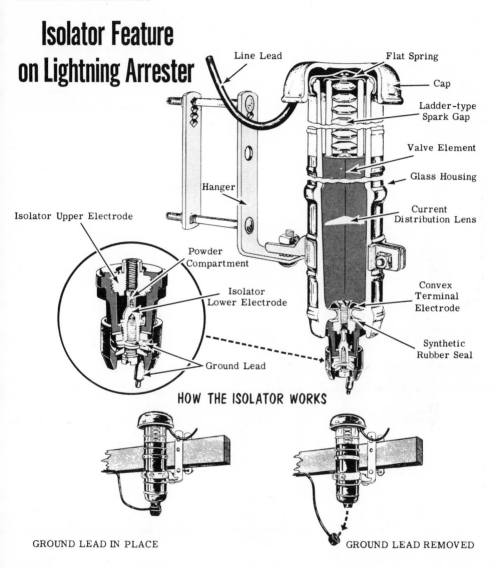

Line Lead

Flat Spring

Cap

Ladder-type Spark Gap

Valve Element

Glass Housing

Current Distribution Lens

Hanger

Isolator Upper Electrode

Powder Compartment

Isolator Lower Electrode

Convex Terminal Electrode

Synthetic Rubber Seal

Ground Lead

HOW THE ISOLATOR WORKS

GROUND LEAD IN PLACE

GROUND LEAD REMOVED

The arrester shown on this page has two advantages: (1) Its glass housing is a cheap insulating container and (2) it is equipped with an isolator which automatically disconnects the arrester from the ground lead in case the unit is electrically damaged. The lineman can easily see this dropped-out isolator from the ground, taking it as an indication that the arrester needs replacing.

Expulsion-type Arresters

In this type arrester there are two series gaps—one external and one internal —which are to be bridged by the high voltage surge. The second gap formed between 2 internal electrodes is actually inside a fiber tube which serves to quench the line energy when it comes through. (Thus the valve element in this case is the fiber tube itself.) When lightning occurs, the gaps are bridged and the high energy flows harmlessly to ground. However, when the line energy tries to go through the arcing channel, the fiber tube creates non-conducting gases which in turn blow the arc and conducting gases out the vent, thus re-establishing a wall of resistance to the line energy.

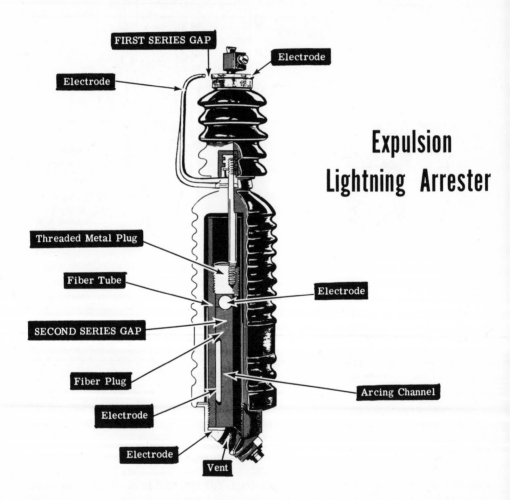

FIRST SERIES GAP

Electrode

Electrode

Expulsion Lightning Arrester

Threaded Metal Plug

Fiber Tube

Electrode

SECOND SERIES GAP

Fiber Plug

Arcing Channel

Electrode

Electrode

Vent

Voltage Regulators

At this point, we are discussing voltage regulators for pole-mounting. Later on when we discuss substations, we will find them used there also.

A voltage regulator is generally used to maintain the voltage of a line. The primary feeder voltage generally drops when a large load current is drawn and less voltage is available across the primaries of the distribution transformers. The regulator maintains the voltage at the proper rated value at all times.

VOLTAGE REGULATOR

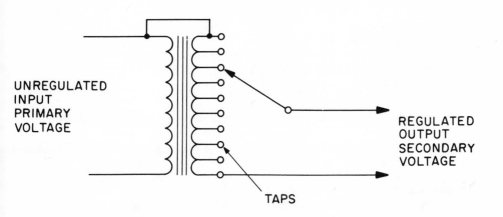

UNREGULATED INPUT PRIMARY VOLTAGE

REGULATED OUTPUT SECONDARY VOLTAGE

TAPS

SCHEMATIC DIAGRAM OF TAP CHANGING UNDER LOAD

The principle of operation of a voltage regulator is somewhat similar to that of a transformer having taps, as previously described. This form of regulator has two fixed windings, a primary (high-voltage) winding connected in shunt or across a line, and a secondary or low-voltage winding connected in series with the line. The secondary or series winding is provided with as many taps as necessary to vary the voltage across this winding. This equipment operates as a voltage regulator by means of a control circuit which automatically changes the tap setting on the series winding, while leaving the voltage applied to the primary (high-voltage) winding alone. The variable voltage in the series winding can thus be added or subtracted from the incoming (or primary) voltage, resulting in an outgoing voltage which can be kept approximately constant even when the incoming primary voltage may vary.

Voltage Regulators (contd.)

An older type, known as the induction type voltage regulator, accomplishes the same effect by having the primary coil rotate, changing its position in relation to the secondary coil, which in this case has no taps.

CUTAWAY VIEW OF LINE-VOLTAGE REGULATOR

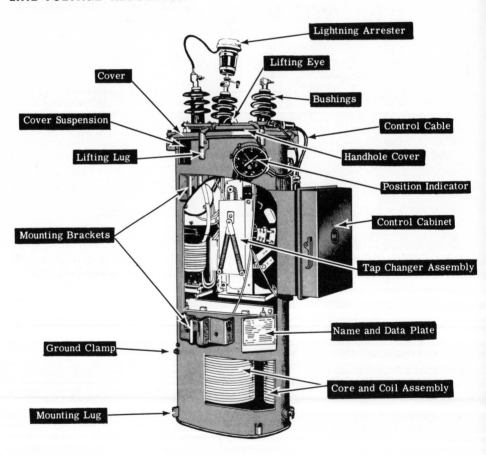

Voltage regulators are either hand or motor-operated. When a motor is used, it is usually automatically controlled by means of relays.

Capacitors

Should the voltage on a circuit fall below a specified level for some reason, a device called a capacitor can momentarily maintain the voltage at line value. Basically, a capacitor serves the same purpose as a storage tank in a water system. By maintaining the water in a storage tank at a definite level, the pressure on the water supplied by the system connected to it is maintained evenly.

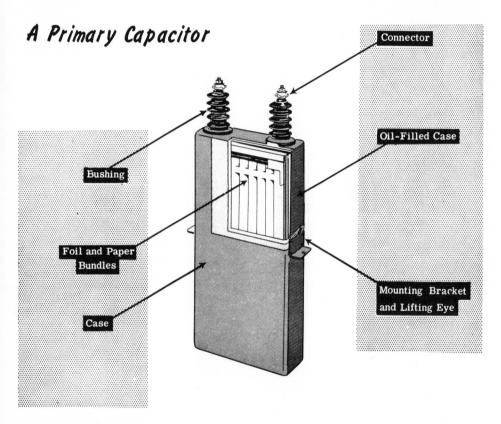

A Primary Capacitor

Connector

Oil-Filled Case

Bushing

Foil and Paper Bundles

Case

Mounting Bracket and Lifting Eye

It is the job of capacitors to keep the *power factor* as close to 1 as possible. The power factor is an important essential of electricity which we will explain in detail later. At this point, let it suffice to say that keeping the power factor close to 1 is a considerable economic advantage to the utility company and to the consumer. *Inductance* is the element in the circuit which is pulling the power factor below 1. *Capacitance* is the enemy of inductance. Therefore, capacitors counteract inductance, keep the power factor close to 1, and save money for the utility company.

Capacitors (contd.)

The capacitor usually consists of two conductors separated by an insulating substance. Among other materials which may be used, a capacitor can be made of aluminum foil separated by oil-impregnated paper, or synthetic insulating materials.

Capacitance is the property of a capacitor. Capacitance depends on the area of the conductors, on the distance between the conductors and on the type of insulating material used.

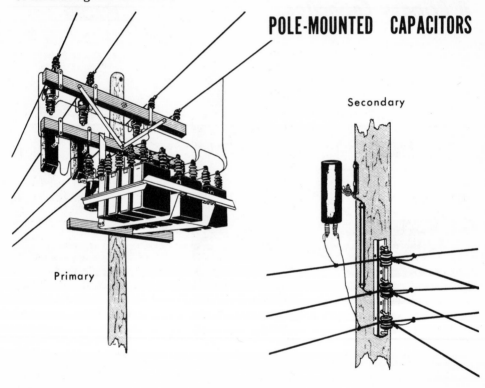

POLE-MOUNTED CAPACITORS

Secondary

Primary

Introducing capacitors into a circuit causes the current to lead the voltage in phase. Introducing inductance (or an inductor) into a circuit causes the current to lag the voltage in phase. In most power applications, inductance prevails and reduces the amount of pay-load power produced by the utility company for a given size of generating equipment. The capacitor counteracts this loss of power and makes power-production more economic.

Capacitors are mounted on crossarms or platforms and are protected with lightning arresters and cutouts, just as transformers.

How Capacitors Are Used

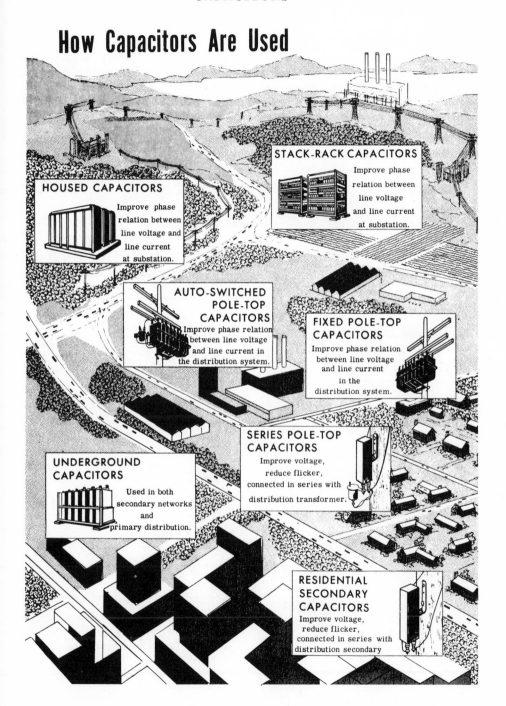

STACK-RACK CAPACITORS
Improve phase relation between line voltage and line current at substation.

HOUSED CAPACITORS
Improve phase relation between line voltage and line current at substation.

AUTO-SWITCHED POLE-TOP CAPACITORS
Improve phase relation between line voltage and line current in the distribution system.

FIXED POLE-TOP CAPACITORS
Improve phase relation between line voltage and line current in the distribution system.

SERIES POLE-TOP CAPACITORS
Improve voltage, reduce flicker, connected in series with distribution transformer.

UNDERGROUND CAPACITORS
Used in both secondary networks and primary distribution.

RESIDENTIAL SECONDARY CAPACITORS
Improve voltage, reduce flicker, connected in series with distribution secondary

(1-79)

Switches

Switches are used to interrupt the continuity of a circuit. They fall into two broad classifications: air switches and oil switches. As their names imply, air switches are those whose contacts are opened in air, while oil switches are those whose contacts are opened under oil. Oil switches are usually necessary only in very high-voltage, high-current circuits.

Air switches are further classified as air-break switches and disconnect switches.

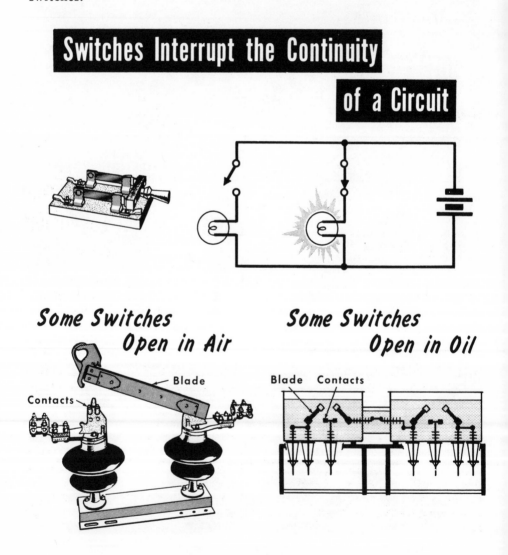

Switches Interrupt the Continuity of a Circuit

Some Switches Open in Air

Blade

Contacts

Some Switches Open in Oil

Blade   Contacts

(1-80)

SWITCHES

Air-break Switches

AIR-BREAK SWITCH IN CLOSED POSITION (WITH ARCING HORNS)

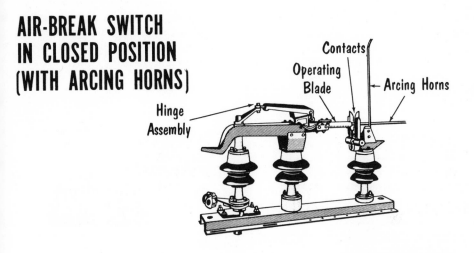

POLE-MOUNTED AIR-BREAK SWITCH IN OPEN POSITION

The air-break switch has both the blade and the contact equipped with *arcing horns*. These are pieces of metal between which the arc resulting from opening a circuit carrying current is allowed to form. As the switch opens, these horns are spread farther and farther apart and the arc is lengthened until it finally breaks.

Air-break switches are of many designs. Some are operated from the ground by a hook on the end of a long insulated stick; some others through a system of linkages are opened by a crank at the foot of the pole. Where more than one conductor is opened, there may be several switches mounted on the same pole. These may be opened singly or altogether in a "gang" as this system is called. Some switches are mounted so that the blade opens downward and these may be provided with latches to keep the knife blade from jarring open.

Air-break Switches (contd.)

A modern development of the air-break switch is the load-break switch which breaks the arc inside a fiber tube. As in the expulsion lightning arrester, the fiber tube produces a gas which helps to confine the arc and blow it out. There is a possibility that the unconfined arc associated with the horn-type switch might communicate itself to adjacent conductors or structures, causing damage and possible injury. But the load-break switch eliminates this hazard.

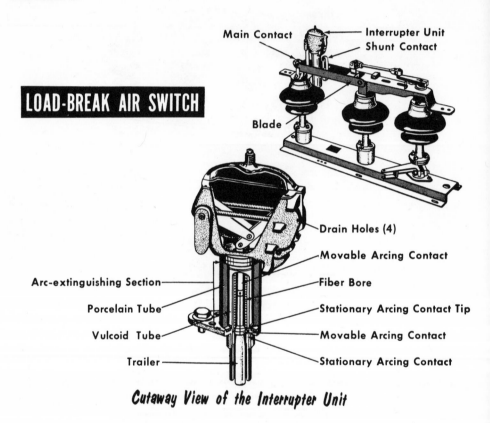

Cutaway View of the Interrupter Unit

The important element of the load-break switch is the interrupter unit shown here. Naturally, the heart of the unit is the arc-extinguishing section which consists of a pair of arcing contacts (one stationary and one movable) and a trailer operating within a fiber bore. The trailer which is made of acrylic resin, follows the contact through the bore, confining the arc between the fiber wall and itself. The arc is extinguished by deionizing gases coming from both the fiber and the acrylic resin.

Disconnect Switches

A DISCONNECT SWITCH

Ring for
Hook-stick
Operation

Contacts

Conductor

Operating
Blade

Mounting
Insulator

The disconnect switch is not equipped with arc-quenching devices and, therefore, should not be used to open circuits carrying current. This disconnect switch isolates one portion of the circuit from another and is not intended to be opened while current is flowing. Air-break switches may be opened under load, but disconnect switches must not be opened until the circuit is interrupted by some other means.

THE DISCONNECT SWITCH IS NEVER OPENED UNDER LOAD

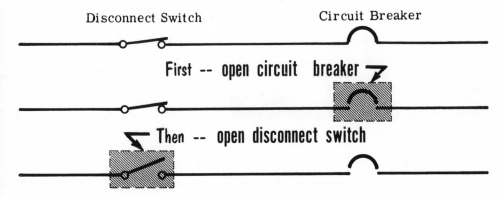

Disconnect Switch   Circuit Breaker

First -- open circuit breaker

Then -- open disconnect switch

SWITCHES

Oil Switches

The oil switch has both the blade and the contact mounted in a tank filled with oil. The switch is usually operated from a handle on the outside of the case. As the switch opens, the arc formed between the blade and contact is quenched by the oil.

An OIL SWITCH Can be Operated Manually or by Remote Control

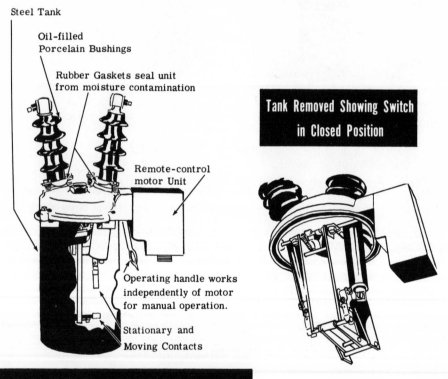

Steel Tank

Oil-filled
Porcelain Bushings

Rubber Gaskets seal unit
from moisture contamination

Remote-control
motor Unit

Operating handle works
independently of motor
for manual operation.

Stationary and
Moving Contacts

Tank Removed Showing Switch in Closed Position

Cutaway View Showing Switch in Open Position

Oil switches may be remote-controlled as well as manually operated. They are used for capacitor switching, street lighting control and automatic disconnect in case of power failure.

Oil Circuit Reclosers

A recloser consists essentially of an oil switch or breaker actuated by relays which cause it to open when predetermined current-values flow through it. The recloser resembles the repeater fuse cutout described previously in many ways. Reclosers are usually connected to protect portions of primary circuits and may take the place of line fuses. The switch or breaker is arranged to reclose after a short interval of time and re-open again should

SINGLE-PHASE OIL-CIRCUIT RECLOSER

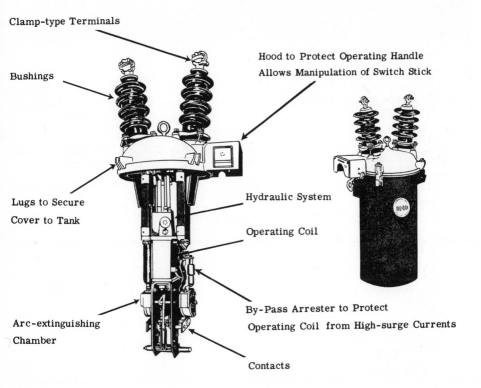

Clamp-type Terminals

Bushings

Hood to Protect Operating Handle
Allows Manipulation of Switch Stick

Lugs to Secure
Cover to Tank

Hydraulic System

Operating Coil

Arc-extinguishing
Chamber

By-Pass Arrester to Protect
Operating Coil from High-surge Currents

Contacts

the fault or overload which caused the excess current-flow persist. Also like the repeater fuse cutout, the recloser can be set for 3 or 4 operations before it locks itself open for manual operation. It differs from the repeater fuse cutout in that its action is repetitive. In a recloser, there is an operating rod actuated by a solenoid plunger which opens and closes the contacts, whereas the repeater fuse works only when the metal has been melted by overheat.

Oil Circuit Reclosers (contd.)

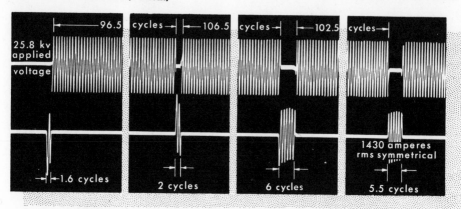

However, should it be desirable to delay the action of the recloser, it can be done by an ingenious timing device. Here is an oscillogram showing a typical example of a recloser operation. Notice that the first time it opens and closes, the action is instantaneous requiring only 1.6 cycles. The second time the action is delayed to 2 cycles, the third time to 6 and the fourth time to 5½ cycles. Then the recloser locks itself open and a lineman must correct the fault and manually close the mechanism.

Distribution transformers are used to step voltage down from the primary main to the secondary main. They can be mounted (1) directly on the pole by lug nuts welded on the transformer case, (2) on a crossarm by hanger irons, or (3) on a platform or pad. The manner of mounting depends on their size and number. The CSP (self-protected) transformer is more easily mounted than the conventional transformer because it has its own primary fuse link and lightning arrester.

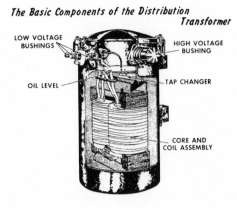

The Basic Components of the Distribution Transformer

LOW VOLTAGE BUSHINGS

HIGH VOLTAGE BUSHING

OIL LEVEL

TAP CHANGER

CORE AND COIL ASSEMBLY

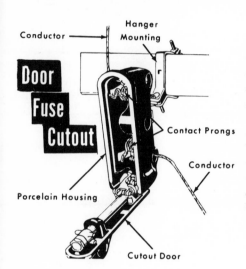

Conductor

Hanger Mounting

Door Fuse Cutout

Contact Prongs

Conductor

Porcelain Housing

Cutout Door

Fuse cutouts are devices which protect the transformer and the primary circuit from dangerous overload. Types of fuse cutouts are (1) door and (2) open. The door-type exudes gas in case of overload. When its door drops open, the lineman knows the fuse is blown. In the open-type cutout, a fiber tube drops when an overload blows the fuse. The repeater cutout is a variation of the automatically reclosing open-type cutout.

The lightning arrester protects the transformer and other equipment from overload. The elementary lightning arrester consists of an air gap in series with a resistive or valve element. When overload occurs, it jumps the air gap and passes through the valve element to ground. This same valve element, however, also prevents the line energy from following the surge to ground.

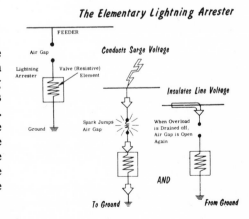

The Elementary Lightning Arrester

FEEDER

Air Gap

Conducts Surge Voltage

Lightning Arrester

Valve (Resistive) Element

Insulates Line Voltage

Ground

Spark Jumps Air Gap

When Overload is Drained off, Air Gap is Open Again

AND

To Ground

From Ground

REVIEW

VOLTAGE REGULATOR

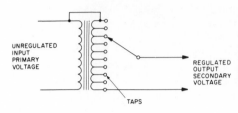

UNREGULATED
INPUT
PRIMARY
VOLTAGE

REGULATED
OUTPUT
SECONDARY
VOLTAGE

TAPS

SCHEMATIC DIAGRAM OF TAP CHANGING UNDER LOAD

A voltage regulator is really a transformer with a variable ratio. It is used to maintain the line voltage at a predetermined value. A capacitor also maintains the line voltage, but does so in relation to the current. By improving the phase relation of the voltage and current on a circuit, a capacitor keeps the power factor close to 1 and saves money.

Switches interrupt the continuity of a circuit. The air-break switch is most commonly used for pole-mounting. Most air-break switches have an operating blade, a set of contacts and arcing horns. The load-break switch has an interrupter unit instead of arcing horns. The disconnect switch is another variation which only isolates one part of a circuit which has been otherwise interrupted.

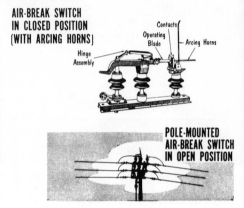

AIR-BREAK SWITCH IN CLOSED POSITION (WITH ARCING HORNS)

Contacts
Operating Blade
Arcing Horns
Hinge Assembly

POLE-MOUNTED AIR-BREAK SWITCH IN OPEN POSITION

Study Questions

1. What is the purpose of a distribution transformer? What are its essential components?

2. Why is the mounting of a distribution transformer important? List some of the methods of doing so.

3. What is the difference between a conventional and a self-protected transformer?

4. What is the function of the fuse cutout?

5. Explain the elementary lightning arrester.

6. List some of the types of lightning arresters and the principles upon which they operate.

7. Why is the capacitor so important to the utility company?

8. What are the two broad classifications of switches?

9. What is the advantage of the "load-break" switch?

10. What is an oil circuit recloser and how does it operate?

Obstacles to Overhead Construction

We have covered some of the basic elements of overhead construction—conductors, insulators, the supports which carry them, transformers, lightning arresters, switches and fuses. Now we shall consider some of the obstacles that the utility company must overcome when constructing its overhead lines. Trees, pedestrians, railroads, rivers, weather and hills all present problems which must be overcome.

Obstacles to be Overcome in Designing and Constructing an Overhead Distribution System

Clearances

In constructing overhead distribution lines, the utility company must consider the terrain as well as man-made obstacles such as railroads. For reasons of safety, the voltage of the wires to be strung must be considered. The higher the voltage, the farther away it must be strung from people, traffic and other wires.

The National Electric Safety Code has set forth specifications governing these clearances. These specifications usually pertain to "class" voltage values. In the following examples, we are using these "class" values rather than actual voltages in use.

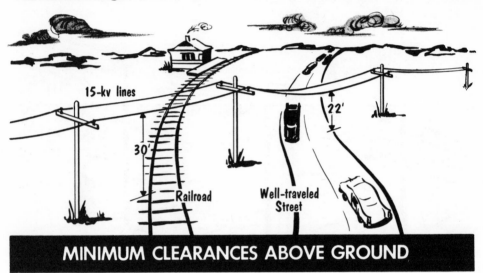

MINIMUM CLEARANCES ABOVE GROUND

Minimum Wire Clearances above Ground or Rails (in feet)

Type of Location	Guys, Messengers, etc.	0 to 750	750 to 15,000	15,000 to 50,000
		Voltages		
When crossing above:				
Railroads	27	27	28	30
Streets, roads, alleys	18	18	20	22
Private driveways	10	10	20	22
Pedestrian walks	15	15	15	17
When wires are along:				
Streets or alleys	18	18	20	22
Roads (rural)	15	15	18	20

Clearances (contd.)

Suppose our 15-kv-distribution line must cross a railroad. According to the specifications given, it must be 30 ft above the ground. In other words, its *minimum clearance* is 30 ft. But then, a few blocks away, it need only cross a well-traveled street. Here the minimum clearance drops to 22 ft.

Sometimes, wires must cross other wires. Here the voltages of both must be considered to make certain that there is no overstrain on the insulation and no flashover. For instance, if a 750-v wire must cross another 750-v wire, there need only be 2 ft between. However if one wire should be 110 kv and the other 8.7 kv, we must allow $6\frac{1}{2}$ ft between.

MINIMUM CLEARANCES BETWEEN WIRES

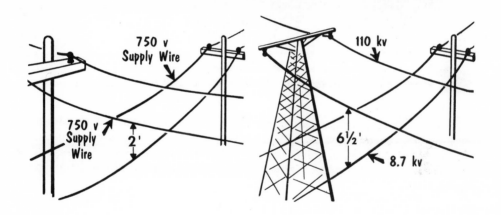

Minimum Clearances above Other Wires (in feet)

Nature of wires crossed over	Commun-ication Guys	Services, Guys, Arrester Grounds	Voltages between Wires 0 to 750	750 to 8700	8700 to 50,000
Communication circuits	2	2	4	4	6
Aerial Supply Cables	4	2	2	2	4
Open Supply Wire, 0 to 750 v	4	2	2	2	4
Open Supply Wire, 750 to 8700 v	4	4	2	2	4
Open Supply Wire, 8700 v to 50,000 v	6	4	4	4	4
Services, guys, arrester grounds	2	2	2	4	4

CLEARANCES

Tree Wires

Trees are a frequently encountered obstacle to distribution overhead wiring. Should a tree come in contact with a primary line, it could act as a ground and short the whole circuit. The abrasion of tree branches can often damage the conductor insulation.

Where possible, trees are trimmed. Permission for trimming must always be obtained either from the owner or from local or state authorities. But permission is not always granted; so a tree wire is necessary.

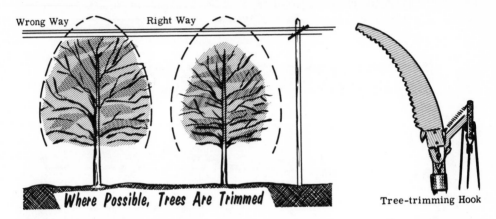

Wrong Way   Right Way

Where Possible, Trees Are Trimmed

Tree-trimming Hook

This wire has a tough, outer covering which can withstand considerable abrasion. Pictured below is a plastic-covered wire which is often used for this purpose. This wire provides partial insulation as well as a tough covering for wires operating at higher-voltage values. There are also other types of tree wire.

Plastic coverings have come to be widely used in the power utility industry because they are highly resistant to the action of moisture, minerals, oils and numerous organic solvents. It must be remembered that such coverings are not to be considered as sufficient insulation for the voltage at which the conductor is operating. The conductor must be handled as if it were bare.

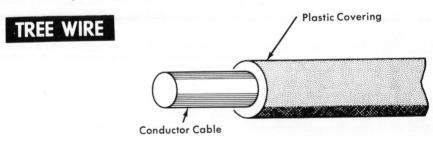

TREE WIRE

Plastic Covering

Conductor Cable

Sag

Notice how a conductor is strung between poles. It is not pulled tight; it sags. In hot weather it sags even more than in cold weather.

Let's consider the analogy of the clothesline again. A tightly strung clothesline puts a much greater strain on the poles which hold it at either end. Adding the weight of wet clothes to a tightly strung line might even pull the poles out of place. However to use a little more rope and allow the clothesline to dip a little would relieve this pull considerably. This dip is what we call sag.

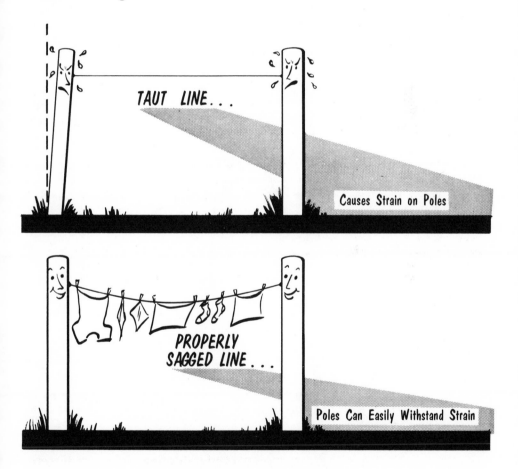

The same applies to stringing wire. When metallic wire is strung tightly, it produces a greater strain on the insulator pins and on the pole. Scientific sag is an important factor in stringing wire.

How the Weather Affects Conductors

We said that a line sags more in hot weather and less in cold weather. This is because conductors expand in hot weather; in other words the length of the conductor increases as the temperature increases. It follows that in cold weather the metallic conductor will be shorter than in warm weather. If the wire were strung without sag, it would snap during cold weather. On the other hand, there is a chance a wire strung with too much sag would dip below the specified clearance in warm weather.

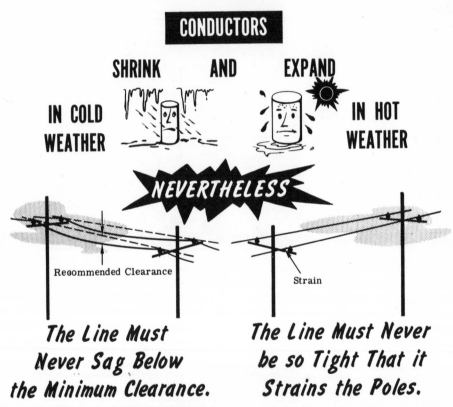

CONDUCTORS

SHRINK **AND** **EXPAND**

IN COLD WEATHER **IN HOT WEATHER**

NEVERTHELESS

Recommended Clearance

Strain

The Line Must Never Sag Below the Minimum Clearance.

The Line Must Never be so Tight That it Strains the Poles.

For copper conductors, the change in length within a temperature range of 100° to 0° is over 5 ft per 1000 ft; for aluminum the change is almost 7 ft.

Besides the temperature, there are other factors which must be considered in determining the sag of a conductor; for example, the length of the span, the weight of the conductors, and wind and ice loading. Also, conductors must not touch each other when swaying in the wind. Tables have been drawn up which consider all these factors and help the utility company determine proper sag.

Guys

Earlier in this book we went into detail about how carefully poles are chosen to carry the load placed on them by the conductors. Careful specifications have been drawn up regarding the length, strength, measurements and setting depth of a pole for every individual situation. In spite of all this care and planning, situations arise where the conductor tries to force the pole from its normal position. This happens because of abnormal loads of ice, sleet, snow and wind as well as because of broken lines, uneven spans, corners, dead-ends and hills.

GUYS ARE USED TO KEEP POLES IN POSITION

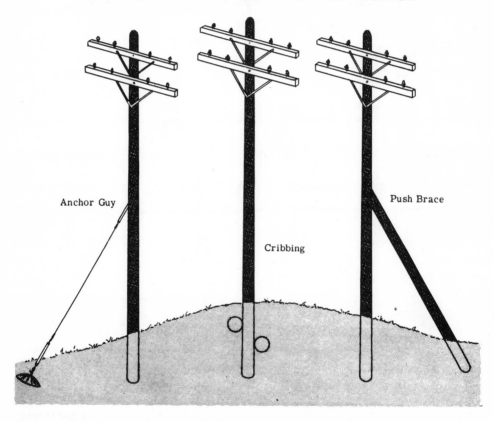

Anchor Guy

Cribbing

Push Brace

When these cases of strain arise, the pole is strengthened and kept in position by guys. Guys are braces or cables fastened to the pole. The most commonly used form of guying is the anchor guy.

Anchor Guys

The major components of an anchor wire guy are the wire, clamps, the anchor, and sometimes a strain insulator. The wire is usually copperweld or galvanized or bethanized steel. The guy is usually firmly attached to the pole by a thimble-eye or by a guy eye bolt and a stubbing washer.

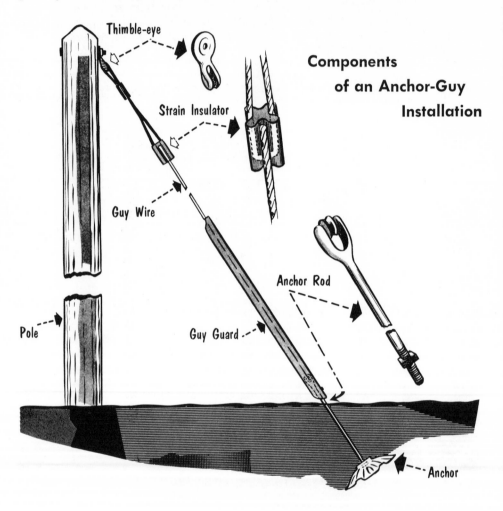

Components of an Anchor-Guy Installation

Thimble-eye
Strain Insulator
Guy Wire
Pole
Guy Guard
Anchor Rod
Anchor

The wire is held firmly by the clamp while the bolt is being tightened. Strain insulators were once commonly used on guy wires; however, they are unnecessary in grounded systems (which are more prevalent).

A guy rod forms a connection between the anchor and the guy wire.

Guy Guards

In the construction of anchor guys, the safety and welfare of the public is of primary concern to the utility company. In well-traveled areas, the part of the guy wire nearest the ground is covered with a protector or guard. This serves a dual purpose. (1) It makes the wire more visible to prevent pedestrians from tripping; (2) Should a person walk into a guy wire, there is no chance of being cut by the wire; (3) Protects the guy wire from damage by cars or vandals.

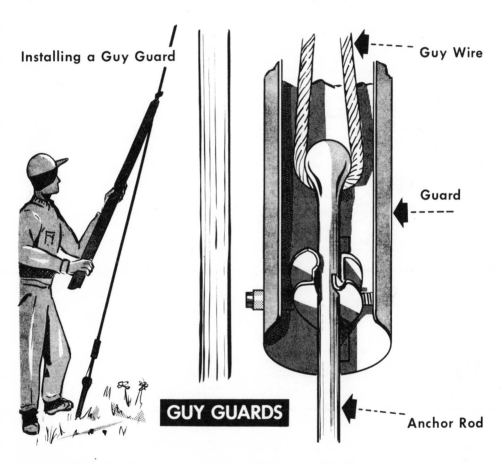

Installing a Guy Guard

Guy Wire

Guard

GUY GUARDS

Anchor Rod

At one time a guy guard consisted chiefly of a round metal tube or some wooden blocks. However, utility companies found that it was a favorite pastime of children to stuff them with paper and junk. So the semi-open guard shown above was developed. Note its rounded edges. This design is also cheaper and easier to install than other types of guy guards.

Anchors

The value of an anchor is determined by its ability to hold the guy wire under strain. At one time, logs (called dead-men) were buried in the ground to anchor the guy wire. Initially, this was a pretty solid anchor. But soil conditions often deteriorated the wood. In any case, the digging of a hole large enough to bury a log is inordinately expensive.

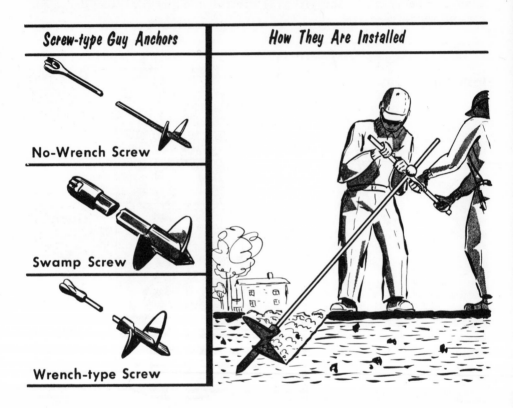

| Screw-type Guy Anchors | How They Are Installed |

No-Wrench Screw

Swamp Screw

Wrench-type Screw

Manufacturers offer a wide variety of anchors—one to suit every type of ground, every particular situation. Anchors can be installed securely in any type of ground from swamp to solid rock.

Sometimes temporary guying is necessary. In other words, there is a chance that the land where the anchor is planted may later become "off-limits" to the utility company. For example, a building may be built there, or the owner may decide to construct a driveway. Whatever the reason, screw type anchors are generally used in temporary situations because they are easily retrievable. They need only be screwed back up and they are ready to be used again. Extra large screw anchors are also used where the soil is swampy or sandy.

GUY ANCHORS AND THEIR INSTALLATION

8-WAY EXPANDED ANCHOR

Designed to Expand to Depicted Size Only After It Has Been Inserted Into the Hole --- Saves Digging.

SINGLE-PLATE ANCHOR

Pulls Against the Solid, Undisturbed Earth --- Not Against Backfill.

CROSS-PLATE ANCHOR

Hole is Bored Straight Down (Often by Machine). Undercut is Dug by Hand.

8-WAY CONE ANCHOR

Used for Clay or Rocky Soils

EXPANDING ROCK ANCHOR

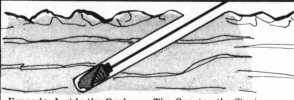

Expands Inside the Rock --- The Greater the Strain, the More Firmly It Is Wedged in the Rock

GUYING POLES

Types of Guys

In situations where an anchor guy is impractical because of crossing a road, a span guy is used. In this form of guying, the guy wire extends from the top of the pole to be guyed to the top of another pole across the street, at approximately the same height. A span guy merely transfers some strain from one pole to another. Hence, a head guy or anchor guy is usually used with a span guy.

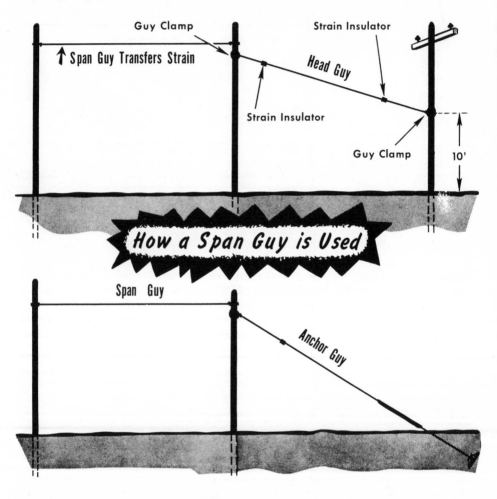

The head guy, is a variation of the span guy, differing in that the wire runs to a point somewhere below the top of the sustaining pole. This type is rarely used in crossing a highway.

Types of Guys (contd.)

Specifications have been set up regarding clearances for guys, just as for conductors.

When it would be difficult to achieve these clearances, a stub guy is used. A stub is just a piece of wood to which the guy wire is attached. The guy must be attached to the stub at some point 8 ft or more above ground. The stub guy wire must allow enough clearance for traffic.

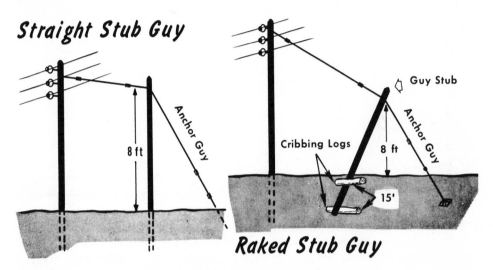

Straight Stub Guy

Raked Stub Guy

Should it prove necessary to guy a pole on private property, some problems may arise. Sometimes the owner objects to a stub or anchor being planted on his land, but he may approve of a guy wire being attached to a tree. This is certainly not recommended practice, because a live tree is a very undependable sustainer. However, there are cases, where there is no other practical solution.

Tree Used as Stub Guy NOT RECOMMENDED PRACTICE

Where Guys Should be Used

When a conductor terminates (dead-ends) on a pole, we attach a guy to the pole to counteract the pull of the conductors. In case of heavy construction where one guy is inadequate, we may also use a guy on the pole *next* to the last one.

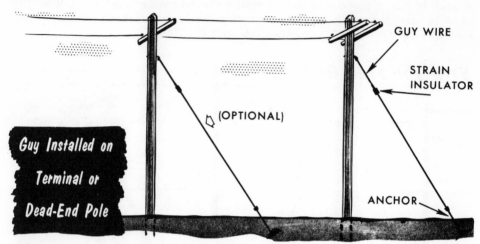

Guy Installed on Terminal or Dead-End Pole

GUY WIRE

STRAIN INSULATOR

(OPTIONAL)

ANCHOR

Sometimes there is the problem of a partial dead-end. In other words, more wires are dead-ended on one side of a crossarm than on the other. A guy is run from the side of the crossarm undergoing the worst strain to the adjacent pole.

Crossarm Guy Compensates Unbalanced Pull on Crossarm

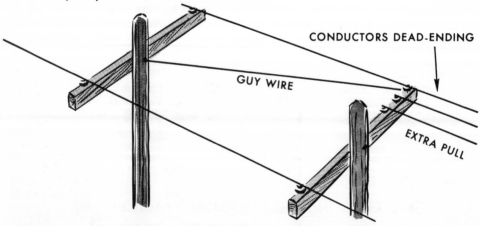

CONDUCTORS DEAD-ENDING

GUY WIRE

EXTRA PULL

GUYING POLES

Where Guys Should be Used (contd.)

Corners are treated like dead-ends; but since the pull on the pole is in two directions, it is wise to use two guys to counteract the pull in two directions.

Two Anchor Guys Sustaining a Corner Pole. Each Guy Counterbalances the Pull of One Set of Wires

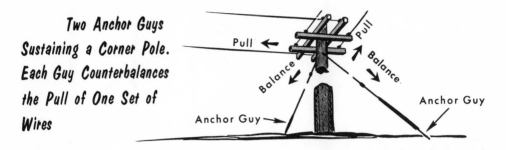

When the conductors form an angle, the pole is submitted to an additional stress. To balance these forces, side guys are attached to the poles to take up the side pull.

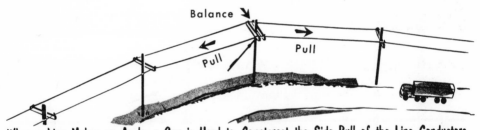

When a Line Makes an Angle, a Guy is Used to Counteract the Side Pull of the Line Conductors.

When a pole is located on the slope of a hill, we can use a head guy or an anchor guy to counteract the down-hill pull of the conductors.

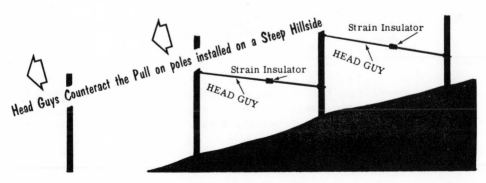

(1-103)

Where Guys Should be Used (contd.)

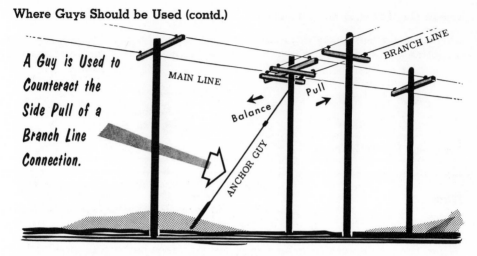

A Guy is Used to Counteract the Side Pull of a Branch Line Connection.

When a branch line shoots from a pole, the weight of the branch conductors causes a side pull. This side pull is counterbalanced by side guys.

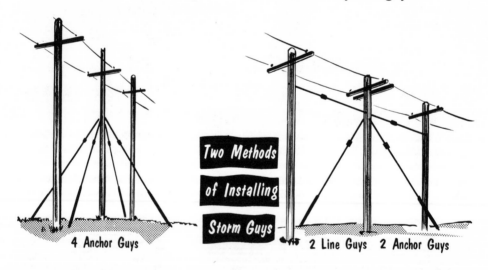

Two Methods of Installing Storm Guys

4 Anchor Guys

2 Line Guys   2 Anchor Guys

When a line may be subjected to storms or other strong atmospheric disturbances, storm guys are placed at regular intervals in the line. Generally, four guys are used for this purpose—two line guys and two side guys. All four guys are fastened to the same pole. Line guys are attached from one pole to the other, thus doing away with the necessity for anchors. Should one pole fail, damage will be limited to the relatively small area between adjacent storm-guyed poles, rather than have the entire line fall, pole after pole.

Other Methods of Sustaining Poles

Sometimes it is necessary to set poles in marshy or swampy land. Since the ground does not hold the pole firmly in such a situation, a guy is necessary. However, it is likely that the use of head or anchor guys might be impractical or impossible for one reason or another. For example, the ground might be too marshy to hold any pole, even a stub; or there might be no available space for an anchor.

Here are some possible solutions:

The pole can be propped up with a push-brace.

The area necessary for guying can be minimized by using a sidewalk guy.

Before a pole is planted in marshy land, an empty oil drum or a tube of corrugated iron is set in the ground. After the pole is dropped in, the drum may be backfilled with dirt or concrete.

Other Methods of Sustaining Poles

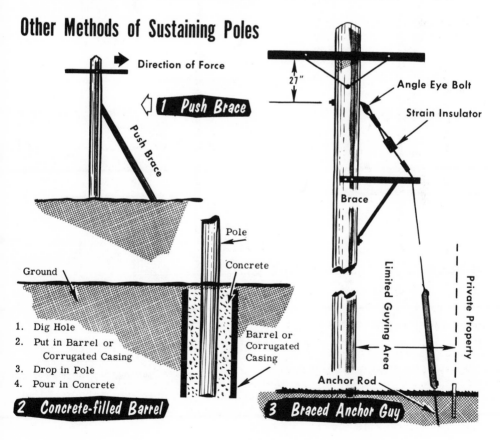

Direction of Force

1 Push Brace

Push Brace

27"

Angle Eye Bolt

Strain Insulator

Brace

Pole

Concrete

Ground

1. Dig Hole
2. Put in Barrel or Corrugated Casing
3. Drop in Pole
4. Pour in Concrete

Barrel or Corrugated Casing

Limited Guying Area

Private Property

Anchor Rod

2 Concrete-filled Barrel

3 Braced Anchor Guy

Other Methods of Sustaining Poles (contd.)

Cribbing is necessary in marshy land when the pole must resist an unbalanced load. It is also used in crowded or residential areas when exposed guys

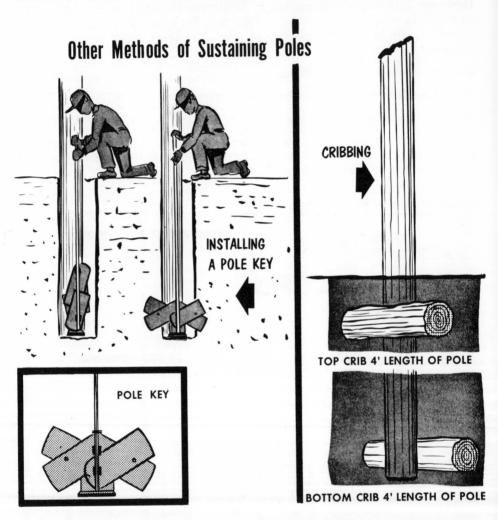

Other Methods of Sustaining Poles

INSTALLING A POLE KEY

POLE KEY

CRIBBING

TOP CRIB 4' LENGTH OF POLE

BOTTOM CRIB 4' LENGTH OF POLE

would be dangerous, unsightly or impractical. Cribbing involves placing logs, stones, or other supports at the bottom of the pole on one side and near the surface of the earth on the other side. These objects counteract the strain being put on the pole from one direction. As in the case of anchors, utility companies have found it economical in some cases to replace logs with another cribbing device called a pole key.

(1-106)

Joint Construction

For economy and appearance, it is often preferable to use one pole for both power and communication lines. This is known as joint construction. Besides the power companies, telephone companies are the biggest users of poles.

Telephone facilities may consist of open wires on crossarms or of a cable supported on a "messenger." Besides the power and telephone companies, the fire department, the police department, the Coast Guard, the railroad, Western Union, cable T.V. systems, and even private citizens may have wires strung on the poles.

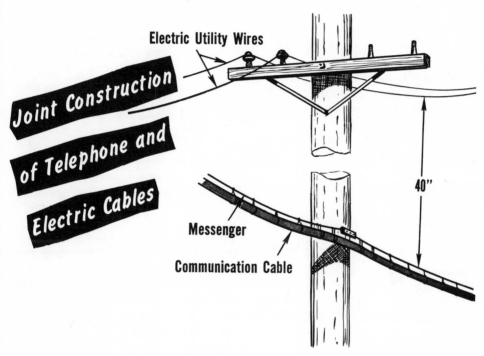

Joint Construction of Telephone and Electric Cables

Electric Utility Wires

40"

Messenger

Communication Cable

All these extra wires add a new factor to the problem of line design. Sometimes extra heavy poles or additional guying become necessary, especially in the case of telephone cables. We must consider carefully the distance between the conductors and equipment of the two utilities. These distances are usually greater than between two power facilities.

Usually the power companies and the communication companies draw up contracts which specify the conditions under which both companies may share the same facilities. Usual specifications include a minimum of 40 inches spacing between these facilities.

MINIMUM CLEARANCES ABOVE GROUND

Conductors must be strung high enough to *clear* railroad crossings, trees, street traffic and other existing wires safely. Specifications have been set up for all these situations, depending on the voltage of the wires. When trees present a clearance problem, they are trimmed. If that is not possible, the conductor is heavily protected with suitable high-strength covering.

Scientific sag is an important factor in stringing wire. The weather affects sag greatly because conductors shrink in cold weather and expand in hot weather. Sag is also affected by the length of the span, the weight of the conductors and wind and ice loadings.

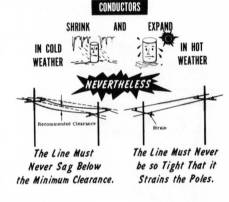

The Line Must Never Sag Below the Minimum Clearance.

The Line Must Never be so Tight That it Strains the Poles.

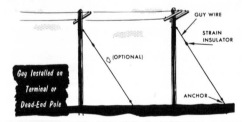

Guy Installed on Terminal or Dead-End Pole

The most commonly used form of guying consists of a wire attached from some point high on a pole to an anchor in the ground. At one time, the anchor was simply a log. Nowadays, mechanical anchors are used. There is a style for every guying need.

When an anchor guy would be impractical, poles can also be sustained by span guys, head guys and stub guys. A span guy simply transfers some of the strain to another pole where it might be more easily balanced with a head, stub or anchor guy. The head and stub guys are constructed on the same principle, that of transferring the strain to a lower point on another pole.

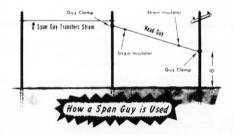

How a Span Guy is Used

REVIEW

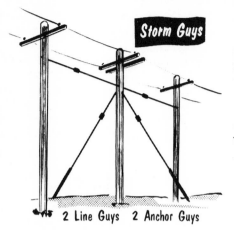

Storm Guys

2 Line Guys   2 Anchor Guys

Guying is necessary in situations of strain such as at dead-ends, corners, angles and hillsides. Guys are also necessary for crossarms which have more wires dead-ended on one side than the other. Should a line come down in a storm, storm guys (each consisting of two line guys and two anchor guys) keep most of the poles upright.

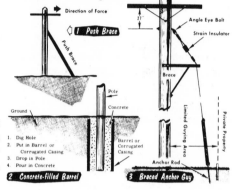

Other Methods of Sustaining Poles

Poles may also be sustained by a push brace, by cribbing or by a concrete foundation. Like log anchors, logs for cribbing have been replaced by mechanical devices.

Study Questions

1. What factors influence the clearances required for overhead lines?

2. What is meant by the *sag* of a conductor?

3. What factors affect a conductor's sag?

4. Why is it necessary for conductors to be properly sagged?

5. What are guys and why are they used?

6. What is the most commonly used form of guying? Name some of its components.

7. Why are guy guards used?

8. List some other ways of sustaining poles.

9. List some situations where guys should be used.

10. What is meant by *joint* construction?

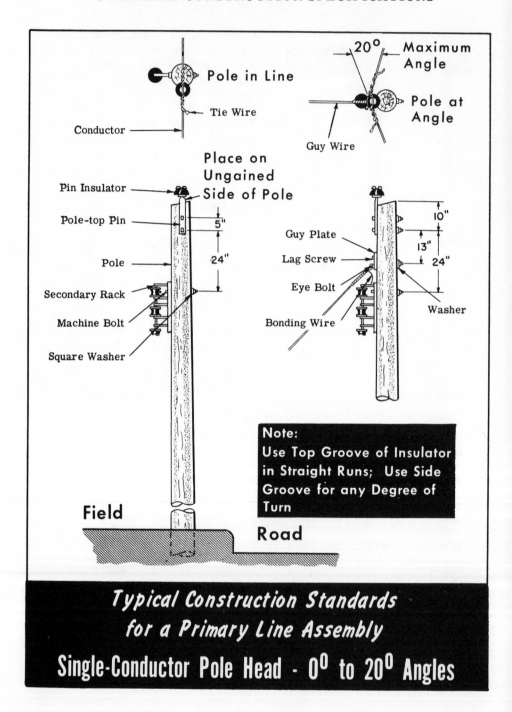

Typical Construction Standards
for a Primary Line Assembly
Single-Conductor Pole Head - 0° to 20° Angles

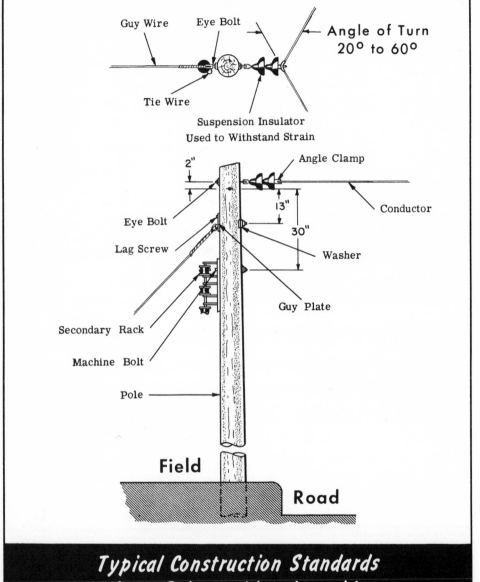

Typical Construction Standards for a Primary Line Assembly

Single-Conductor Pole Head - 20⁰ to 60⁰ Angles

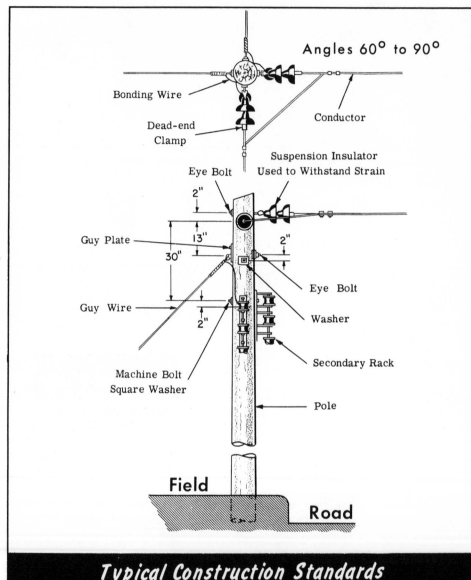

Angles 60° to 90°

Bonding Wire

Conductor

Dead-end Clamp

Suspension Insulator Used to Withstand Strain

Eye Bolt

2"

Guy Plate

13"

2"

30"

Eye Bolt

Washer

Guy Wire

2"

Secondary Rack

Machine Bolt Square Washer

Pole

Field

Road

Typical Construction Standards for a Primary Line Assembly

Single-Conductor Pole - 60⁰ to 90⁰ Angles

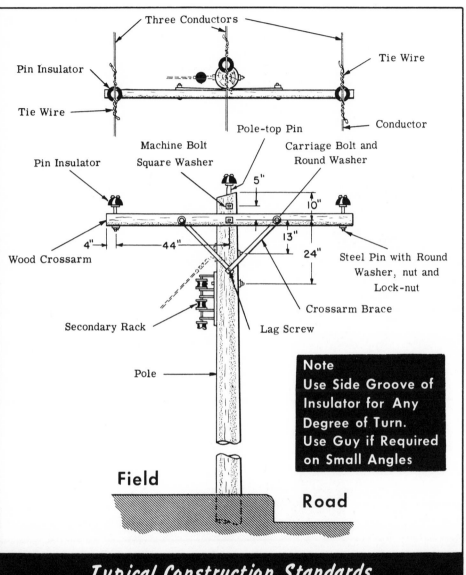

Three Conductors

Tie Wire

Pin Insulator

Tie Wire

Conductor

Pole-top Pin

Machine Bolt
Square Washer

Carriage Bolt and
Round Washer

Pin Insulator

5"

10"

Wood Crossarm

4"

44"

13"

24"

Steel Pin with Round
Washer, nut and
Lock-nut

Crossarm Brace

Secondary Rack

Lag Screw

Pole

Note
Use Side Groove of
Insulator for Any
Degree of Turn.
Use Guy if Required
on Small Angles

Field

Road

Typical Construction Standards
for a Primary Line Assembly

Three-Conductor Pole Head -0⁰ and Small Angles

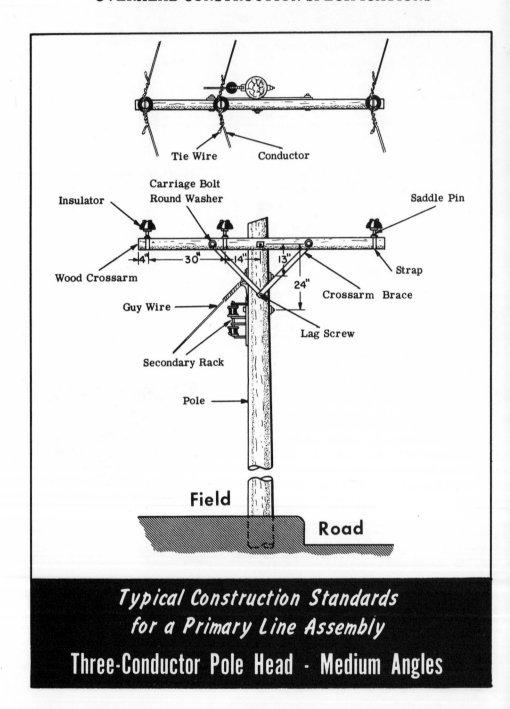

Tie Wire   Conductor

Carriage Bolt
Round Washer

Insulator   Saddle Pin

Wood Crossarm   4"   30"   14"   13"   Strap

Crossarm Brace

Guy Wire   24"

Lag Screw

Secondary Rack

Pole

Field   Road

Typical Construction Standards
for a Primary Line Assembly
Three-Conductor Pole Head · Medium Angles

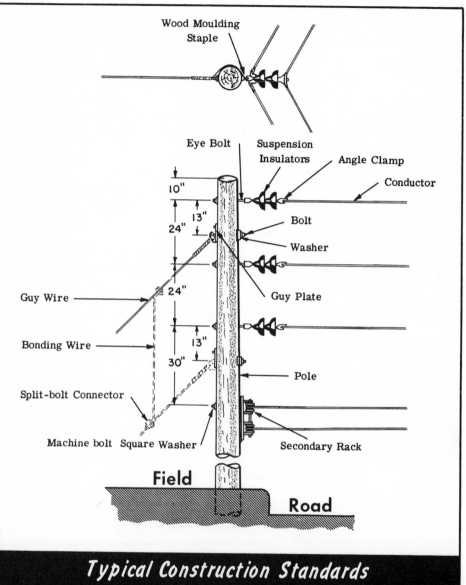

Typical Construction Standards
for a Primary Line Assembly

Three-Conductor Pole Head - Angles up to 60°

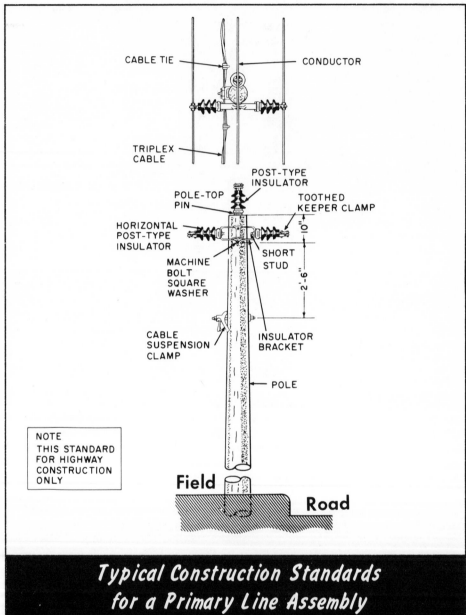

CABLE TIE

CONDUCTOR

TRIPLEX CABLE

POST-TYPE INSULATOR

POLE-TOP PIN

TOOTHED KEEPER CLAMP

HORIZONTAL POST-TYPE INSULATOR

10"

MACHINE BOLT SQUARE WASHER

SHORT STUD

2'-6"

CABLE SUSPENSION CLAMP

INSULATOR BRACKET

POLE

NOTE
THIS STANDARD FOR HIGHWAY CONSTRUCTION ONLY

Field

Road

Typical Construction Standards for a Primary Line Assembly

Three-Conductor Pole Head — 0° to 15° Angles

INDEX

Air-brake switches, 1-55, 1-80 to 1-83, 1-88
Aluminum conductors, 1-12, 1-43 to 1-46,
  1-50, 1-94
Aluminum conductors steel reinforced
  (ACSR), 1-45
Alumoweld conductors, 1-46
American Wire Gage (AWG), 1-49 to 1-50
Anchor guys, 1-96 to 1-99, *see also* Guys
Angle ties, 1-54
Annealed conductors, 1-44
Arcing horns, 1-81, 1-88
Arresters, *see* Lightning arresters
Automatic line splices, 1-52

Braces, push, 1-95, 1-105
Brown & Sharpe Gage, 1-49
Bug nut (split-belt), connectors, 1-52
Bushings, 1-48, 1-56 to 1-57

Cables, 1-13 to 1-14, 1-16
  clearances, 1-90 to 1-92
  joints, 1-107
  obstacles, 1-89
  sag, 1-93
  secondary, 1-13
  weather effects on, 1-94
  see also Conductors
Capacitance, 1-77
Capacitor bushings, 1-57
Capacitors, 1-55, 1-77 to 1-79, 1-88
Ceder poles, 1-9 to 1-20, 1-26
Chestnut poles, 1-19, 1-26
Circuit breakers, 1-64
Circuits, distribution, 1-8
Circular mils, 1-50
Clamps, parallel, 1-52
Clearance, 1-90 to 1-92, 1-101, 1-107 to 1-108
  joint construction, 1-107
  minimum above ground, 1-90 to 1-91
  minimum above other wires, 1-91
Clevis, 1-42
Coil assembly, distribution transformers, 1-56
Concrete poles, 1-17 to 1-18
Conductor connectors, 1-52
Conductor coverings, 1-48 to 1-49

Conductor sizes, 1-49 to 1-50
Conductor stranding, 1-47
Conductor tie wires, 1-51, 1-54
Conductors
  aluminum, 1-12, 1-43 to 1-46, 1-50, 1-94
  alumoweld, 1-46
  annealed, 1-44
  characteristics, 1-50
  copper, 1-12, 1-43 to 1-44, 1-46, 1-50, 1-94
  copperweld, 1-46
  line, 1-43, 1-52
  silver, 1-43
  steel, 1-43, 1-46
Conduits, underground, 1-10
Cone anchors, 1-99
Connectors, 1-52
Copper conductors, 1-12, 1-43 to 1-44, 1-46,
  1-50, 1-94
Copperweld conductors, 1-46
Core assembly, distribution transformers,
  1-56
Cribbing, 1-95, 1-106, 1-109
Crossarms
  plastic, 1-18
  steel, 1-17
  wood, 1-30
Cross-plate anchors, 1-99
CSP transformers, 1-63 to 1-64, 1-87
Current, 1-9
Cutouts, 1-55
  door, 1-87
  open, 1-87
  repeater, 1-87

Dead-ends
  guying, 1-102
  insulation, 1-41
Dead-men, 1-98
Disconnect switches, 1-83, 1-88
Distribution, difference from transmission,
  1-4
Distribution lines
  primary, 1-4
  secondary, 1-4
  sub-transmission, 1-4
  transmission, 1-4

INDEX

Distribution substations, 1-6, 1-10
Distribution systems, 1-8, 1-10 to 1-16
Distribution transformers, 1-6, 1-10 to 1-11,
    1-15 to 1-16
  bushings, 1-57
  conventional, 1-63
  CSP, 1-63 to 1-64, 1-87
  junction, 1-56
  pad mounted, 1-62
  platform mounted, 1-62
  pole mounted, 1-8, 1-59 to 1-62
  tap changers, 1-56, 1-58, 1-75
Distribution voltages
  determination, 1-12 to 1-14
  in use, 1-14
Door fuses, 1-65 to 1-67, 1-87
Double arms, 1-28

Economics
  of transmission and distribution systems,
      1-12 to 1-13, 1-16
  of utility companies, 1-5
Electric current, 1-9
Electrical energy
  consumption, 1-2, 1-7
  distribution cost, 1-12
  generation, 1-1, 1-3 to 1-4, 1-6
  sources, 1-3, 1-7
  transmission, 1-1
Expanding rock anchors, 1-99
Expulsion lightning arresters, 1-74

Federal Power Commission (FPC), 1-7
Feeder lines, 1-10, 1-15
Flat roofing, 1-29
Fossil fuel energy, 1-3, 1-7
Fuels, 1-7
Fuse links, protective, 1-64
Fuses, 1-55, 1-65 to 1-66
  door, 1-67, 1-87
  open, 1-68, 1-87
  repeater, 1-69, 1-87

Gable roofing, 1-29
Gage, wire, 1-49 to 1-50
Gage numbers, 1-49
Gaining of poles, 1-28
Galvanizing, 1-46
Gang switching, 1-81
Generators, 1-1, 1-3, 1-5, 1-7

Glass insulation, 1-37
  lightning arresters, 1-73
Guy guards, 1-96 to 1-97
Guys
  branch lines, 1-104
  clearance, 1-101
  eye bolts, 1-96
  head, 1-100, 1-105, 1-108
  installation, 1-99
  placement, 1-102 to 1-104
  side, 1-104
  sidewalk, 1-105
  span, 1-100, 1-108
  stub, 1-101, 1-105, 1-108
  storm, 1-104, 1-109

Hanger irons, 1-61, 1-87
Hanging insulators, 1-39
Hard drawn conductors, 1-44
Head guys, 1-100, 1-105, 1-108
High-voltage bushings, 1-56
Hotsticks, 1-51
Hydro-electric energy, 1-3, 1-7

Ice loadings, 1-22, 1-24
Inductance, 1-77
Induction voltage regulators, 1-76
Insulators
  ball and socket, 1-40
  clamp top, 1-51
  clevis, 1-42
  petticoats, 1-38
  pin, 1-31, 1-37 to 1-39, 1-110
  post, 1-37 to 1-39
  side groove, 1-54
  spool, 1-39, 1-42
  strain, 1-39, 1-41, 1-96, 1-100, 1-103
  string, 1-40
  suspension, 1-39 to 1-41
Interrupter units, 1-82, 1-88
Isolator lightning arresters, 1-73

Joint construction, 1-107
Junction poles, 1-62

Keeper clamps, toothed, 1-116
Kilowatt, 1-2 to 1-3

INDEX

Lightning arresters, 1-55
  elementary, 1-71
  expulsion, 1-74
  isolator, 1-73
  pellet, 1-72
  valve element, 1-71 to 1-72, 1-87
Line conductors, 1-43 to 1-54
Line equipment, 1-55 to 1-58
Line guys, 1-104, 1-109
Load-brake air switches, 1-82, 1-88
Loading districts, 1-24
Low-voltage bushings, 1-24
Lug nuts, 1-87

Mains, 1-10 to 1-11, 1-15
Mechanical connectors, 1-52
Messenger cables (telephone), 1-107
Meters, watt-hour, 1-11, 1-15
Mounting equipment, 1-59 to 1-62

National Electrical Safety Code (NESC),
  1-98
No-wrench screw anchors, 1-98

Obstacles to overhead construction, 1-89
Oil circuit reclosers, 1-55, 1-85 to 1-86
Oil filled bushings, 1-57
Oil switches, 1-55, 1-80, 1-84
Open fuses, 1-68, 1-87
Oscillograms, 1-86
Overhead construction
  clearances, 1-90 to 1-94
  joints for, 1-107
  obstacles to, 1-89
  specifications, 1-110 to 1-116
  supports for, 1-17

Parallel clamps, 1-52
PDL (Primary Distribution Lines), 1-4
Pellet lightning arresters, 1-72
Petticoats, 1-38
Pine poles, 1-20, 1-26, 1-30
Pin insulators, 1-31, 1-38 to 1-39, 1-110
Pin spacing, 1-32
Plastic poles, 1-38
Plate anchors, 1-99
Pole pins, 1-31 to 1-32, 1-110
Poles
  accessories, 1-30 to 1-34

classification, 1-25 to 1-26
concrete, 1-17 to 1-18
cribbing, 1-106
depth, 1-27
dimensions, 1-26
gaining, 1-28
guying, 1-95 to 1-107
key, 1-106
length, 1-21
plastic, 1-38
roofing, 1-29
spacing, 1-32
standards, 1-21 to 1-29
strength, 1-32
wooden, 1-18 to 1-29
Porcelain bushings, 1-57
Porcelain insulators, 1-37
Post insulators, 1-37 to 1-39
Power factors, 1-77
Pressure voltage, 1-10, 1-12, 1-16
Primary bushings, 1-48, 1-57
Primary Distribution Lines (PDL), 1-4
Primary feeders, 1-10, 1-15
Protective devices, 1-64 to 1-74
Push braces, 1-95, 1-105, 1-109

Racks, secondary, 1-34, 1-110 to 1-115
Raked stub guys, 1-101
Reclosers, oil circuit, 1-55, 1-85 to 1-86
Regulators, voltage, 1-55, 1-75 to 1-76, 1-88
Relays, 1-76
Repeater fuses, 1-69, 1-87
Roofing of poles, 1-29
Rubber safety accessories, 1-48

Saddle pins, 1-114
Safety, 1-12, 1-14, 1-24, 1-32, 1-48, 1-51, 1-90,
  1-97
Sag, 1-93 to 1-94, 1-108
  effects of weather, 1-94
Screw anchors, 1-98
Secondary bushings, 1-57
Secondary cable, 1-33
Secondary distribution, 1-4, 1-8, 1-10 to 1-11,
  1-15
Secondary racks, 1-34, 1-110 to 1-115
Service wires, 1-33 to 1-34
Sheaths, 1-13
Shoes, 1-51
Side guys, 1-104
Sidewalk guys, 1-105

INDEX

Silicon carbide lightning arresters, 1-71
Silver conductors, 1-43
Slant roofing, 1-29
Span guys, 1-100, 1-108
Splices, automatic line, 1-52
Split-belt connectors, 1-52
Spool insulators, 1-39, 1-42
Stations, switching, 1-7 to 1-8, 1-10, 1-15
Steel conductors, 1-43, 1-46
Steel towers, 1-17
Storm guys, 1-104, 1-109
Straight stub guys, 1-101
Strain insulators, 1-39, 1-41, 1-96, 1-100, 1-103
Stranded conductors, 1-47
String insulators, 1-40
Stub guys, 1-101, 1-105, 1-108
Stubbing washers, 1-96
Substations, 1-5 to 1-6, 1-10, 1-15
Sub-transmission lines, 1-4, 1-10
Subway vaults, 1-6
Supports, overhead construction, 1-17 to 1-36
Suspension insulators, 1-39 to 1-40
Swamp screw anchors, 1-98
Switches, 1-12, 1-55
    air-brake, 1-80 to 1-82
    disconnect, 1-83
    interrupter, 1-88
    oil, 1-80, 1-84
    load-breaker, 1-82 to 1-83
Switching stations, 1-7 to 1-8, 1-10, 1-15

Tap changers, 1-56, 1-58, 1-75
Thimble-eye, 1-96
Tie wires, 1-51
Towers, steel, 1-17
Transformers, distribution, *see*
    Distribution transformers
Transmission and distribution systems, 1-1 to 1-16

Transmission lines, 1-4
Tree wires, 1-92
Trees, trimming, 1-92

U. S. Department of Commerce, 1-24
Underground systems, 1-7, 1-10, 1-12 to 1-13
Utility companies
    economics, 1-5, 1-12
    growth, 1-1, 1-3
    organization, 1-5
    service, 1-2

Valve elements, 1-71, 1-87
Valve lightning arresters, 1-72
Vaults, subway, 1-6
Voltage, 1-10, 1-16
    determining distribution, 1-12 to 1-14
    water-current analogy, 1-9, 1-12
Voltage regulators, 1-75 to 1-76, 1-88

Washers, stubbing, 1-96
Watts, 1-2 to 1-3
Watt-hour meters, 1-11, 1-15
Weather, effects on sag, 1-94
Wind loadings, 1-22, 1-24
Winding ratio, 1-58
Wire gages, 1-49 to 1-50
Wire sizes, 1-49 to 1-50
Wood crossarms, 1-30
Wood pole pins, 1-31 to 1-32
Wood poles, 1-20 to 1-29
    dimensions, 1-26
    length, 1-21
    steeing depth, 1-27
    strength, 1-22 to 1-25
    types, 1-20
Wrench screw anchors, 1-98